What Every Engineer Should Know About
Quality Control

WHAT EVERY ENGINEER SHOULD KNOW

A Series

Editor

William H. Middendorf

Department of Electrical and Computer Engineering
University of Cincinnati
Cincinnati, Ohio

Vol. 1 What Every Engineer Should Know About Patents, *William G. Konold, Bruce Tittel, Donald F. Frei, and David S. Stallard*

Vol. 2 What Every Engineer Should Know About Product Liability, *James F. Thorpe and William H. Middendorf*

Vol. 3 What Every Engineer Should Know About Microcomputers: Hardware/ Software Design: A Step-by-Step Example, *William S. Bennett and Carl F. Evert, Jr.*

Vol. 4 What Every Engineer Should Know About Economic Decision Analysis, *Dean S. Shupe*

Vol. 5 What Every Engineer Should Know About Human Resources Management, *Desmond D. Martin and Richard L. Shell*

Vol. 6 What Every Engineer Should Know About Manufacturing Cost Estimating, *Eric M. Malstrom*

Vol. 7 What Every Engineer Should Know About Inventing, *William H. Middendorf*

Vol. 8 What Every Engineer Should Know About Technology Transfer and Innovation, *Louis N. Mogavero and Robert S. Shane*

Vol. 9 What Every Engineer Should Know About Project Management, *Arnold M. Ruskin and W. Eugene Estes*

Vol. 10 What Every Engineer Should Know About Computer-Aided Design and Computer-Aided Manufacturing: The CAD/CAM Revolution, *John K. Krouse*

Vol. 11 What Every Engineer Should Know About Robots, *Maurice I. Zeldman*

Vol. 12 What Every Engineer Should Know About Microcomputer Systems Design and Debugging, *Bill Wray and Bill Crawford*

Vol. 13 What Every Engineer Should Know About Engineering Information Resources, *Margaret T. Schenk and James K. Webster*

Vol. 14 What Every Engineer Should Know About Microcomputer Program Design, *Keith R. Wehmeyer*

Vol. 15 What Every Engineer Should Know About Computer Modeling and Simulation, *Don M. Ingels*

Vol. 16 What Every Engineer Should Know About Engineering Workstations, *Justin E. Harlow III*

Vol. 17 What Every Engineer Should Know About Practical CAD/CAM Applications, *John Stark*

Vol. 18 What Every Engineer Should Know About Threaded Fasteners: Materials and Design, *Alexander Blake*

Vol. 19 What Every Engineer Should Know About Data Communications, *Carl Stephen Clifton*

Vol. 20 What Every Engineer Should Know About Material and Component Failure, Failure Analysis, and Litigation, *Lawrence E. Murr*

Vol. 21 What Every Engineer Should Know About Corrosion, *Philip Schweitzer*

Vol. 22 What Every Engineer Should Know About Lasers, *D. C. Winburn*

Vol. 23 What Every Engineer Should Know About Finite Element Analysis, *edited by John R. Brauer*

Vol. 24 What Every Engineer Should Know About Patents, Second Edition, *William G. Konold, Bruce Tittel, Donald F. Frei, and David S. Stallard*

Vol. 25 What Every Engineer Should Know About Electronic Communications Systems, *L. R. McKay*

Vol. 26 What Every Engineer Should Know About Quality Control, *Thomas Pyzdek*

Other volumes in preparation

What Every Engineer Should Know About
Quality Control

Thomas Pyzdek

Quality America, Inc.
Tucson, Arizona

Marcel Dekker, Inc. New York and Basel
ASQC Quality Press Milwaukee, Wisconsin

Library of Congress Cataloging-in-Publication Data

Pyzdek, Thomas
What every engineer should know about quality control
(What every engineer should know ; vol. 24)
Bibliography: p.
1. Quality control. I. Title. II. Series: What every
engineer should know ; v. 24.
TS156.P99 1988 658.5'62 88-30940
ISBN 0-8247-7966-5

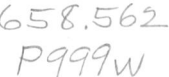

MARCEL DEKKER, INC.
270 Madison Avenue, New York, New York 10016

Current printing (last digit):
10 9 8 7 6 5 4 3 2 1

PRINTED IN THE UNITED STATES OF AMERICA

To my wife, Carol, who put up with me through it all.

Preface

Quality has always been an issue of importance. Many experts believe that the failure to maintain high levels of quality is largely responsible for America's declining position in world markets. The *Harvard Business Review* concluded that if America does not dramatically improve quality "our whole economy comes into jeopardy." In a later issue of the same journal, Arnold O. Putnam calls for a complete redesign of the relationship between engineering and quality, stating that "the venerable American tradition is to design first and do manufacturing engineering and quality control later. This mode of operation has serious drawbacks." Clearly, America must improve quality dramatically and quickly. The engineering community plays a vital role in this effort. It follows that engineering must develop a much better understanding of quality.

Recent developments have made quality much more a science than an art form. The body of knowledge related to quality is now well defined, and it is becoming increasingly sophisticated. Today's issues of consumerism, product and professional liability, and government regulation make formal consideration of quality an absolute necessity.

When the University of Nebraska asked its engineering alumni to name an area where they felt additional training should have been provided, quality was named more than any other area.

Quality is more than simply "goodness." Quality control can be thought of, broadly, as the science of discovering and controlling variations. It is a body of specialized knowledge that encompasses disciplines as diverse as mathematics and management, psychology and engineering, the law and human relations. A central theme of quality control is that well-engineered designs should be consistently reproduced. The importance of this to every engineer is obvious.

The *What Every Engineer Should Know* series tackles a formidable task. As an advertising flier for the series points out, *What Every Engineer Should Know* amounts to a bewildering array of information. In writing this volume on quality control, I am acutely aware of this "bewildering array." The task of selecting the vital facts about quality control that every engineer should know was arduous and often perplexing. In my opinion, as a self-confessed quality fanatic, every engineer should be an *expert* on quality. Obviously, accomplishing this is not possible in a single book. Instead you will receive a sampling from the most important areas—enough to be conversant with the subject and, I hope, enough to stimulate you to learn more. In short, I have opted for breadth over depth.

This book presents the essential elements from the vast body of knowledge of quality control. The reader will learn about the most important quality management and quality engineering methods and tools. By understanding the science of quality, engineers will better appreciate their place in the overall picture. My intent is to provide an overview of the most important elements, occasionally in enough detail to give you a "feel" for an especially important technique (such as control charts). In the future, quality will become ever more important, and the engineers will be asked for their contribution to quality. The ability to make a contribution will depend on how well the basics are understood.

This is not a formal book about quality; there are plenty of those already and I doubt that I can improve on what already exists. One of my objectives in writing this book was to convey the fact that quality is a peculiar mixture of the subjective and the quantitative. A glance at the table of contents will reveal an eclectic array of subjects, all of which play an important role in quality. Still, in spite of almost six decades of effort to make it a science, quality remains somewhat a matter of opinion. We speak of "quality gurus" and their "philosophies" on quality, something unheard of in, say, chemistry or civil engineering. Some of the opinions of the gurus are presented in this book and, for better or for worse, at times I have added my opinion to this chorus. I hope that by conveying some of my personal experiences I can help you gain a better understanding of the process that is "quality control."

Another objective was to give enough information to provide the essence of what quality control involves. While the contents of the book approximately match the body of knowledge required of Certified Quality Engineers (CQEs) by the American Society for Quality Control (ASQC), the depth of treatment of each subject is necessarily much less than that required of CQEs. The book provides a brief overview of the disciplines affecting quality control, with enough detail to give an accurate feel for how the discipline is actually applied to quality control work.

Also included are chapter summaries that tell, among other things, just what material has been omitted; a list of recommended reading that contains additional material; and a bibliography of references cited in the book. You are urged to consult these other works and to continue your study of the vital area of quality control.

I have not spared you my opinions. I believe that most books on quality are too sterile and academic. My years of work in the field have been educational, interesting, and exciting, and I wouldn't want to deprive you of the insights I've acquired along the way! I hope you find my editorial digressions enlightening, entertaining, and educational. If this book succeeds in piquing your interest in quality control, it will have accomplished its goal.

Thomas Pyzdek

Contents

What Every Engineer Should Know About

Quality Control

1

Basic Concepts of Quality Control

WHAT IS QUALITY?

Webster's New 20th Century Dictionary defines *quality* as "the degree of excellence that a thing possesses." Unfortunately, while this is the common usage of the term, it is useless as a working definition. If you were given a set of fine china, chances are you'd agree that the dinnerware was "quality merchandise." But what if you were asked to judge the quality of a set of paper plates? Although it is obvious that the paper plates don't compare to fine china, it's equally clear that the users of these products have certain things in mind when they make their purchase. The fine china would be as out of place on a picnic as the paper plates would be at a formal dinner. Thus, if we want a definition of quality that works with paper plates as well as with fine china, we must look beyond the dictionary.

The quality profession has struggled with the definition of quality for quite some time. Some of the most respected people in the quality profession have advanced their own definitions. Dr. W.Edwards Deming defines two different types of quality. Quality of *conformance* is the extent to which a firm and its suppliers surpass the design specifications required to meet the customer's needs. Quality of *performance* is the measure, determined through research and sales/service call analysis, of how well products perform in the marketplace. This leads to redesign, new specifications, and never-ending improvement.

Other quality experts have other definitions. Philip Crosby says quality is conformance to requirements, period. Joseph Juran defines quality as fitness for use. To Armand Feigenbaum, quality is "the total composite product and service characteristics of marketing, engineering, manufacture, and maintenance through which the product and service in use will meet the expectations of the customer." Unlike the dictionary definition, these definitions will provide a basis for developing a quality control system. It remains, however, for us to determine what is meant by "control."

WHAT IS CONTROL?

Once again, let's start with Webster. One definition of control is "a standard of comparison for verifying or checking the findings of an experiment." This time Webster comes quite close to providing a definition we can work with, as can be seen by comparing this to Juran's definition of control: "The process of measuring quality performance, comparing it to requirements, and acting on the difference."

Feigenbaum defines control as "a process for delegating responsibility and authority for a management activity while retaining the means of assuring satisfactory results." Feigenbaum's definition is sufficiently generic to apply to any activity, not just control of quality. This can be seen even more clearly if we look at Feigenbaum's four steps of control:

1. Setting standards
2. Appraising conformance
3. Acting when necessary
4. Planning for improvements

THE QUALITY CONTROL SYSTEM

At one time quality control was viewed, quite narrowly, as product inspection. "Quality control" consisted of assigning the last person on the assembly line the task of ensuring that the product worked. In modern practice, quality control begins with the design process and continues through manufacturing and use of the product. The sum of all of these efforts is called *total quality control (TQC)*. Quality control can also be viewed as all activities directed toward discovering and controlling variation in performance. The principles of quality control can be applied equally well to either products or services. We always seek, as an ultimate goal, products and services of consistent excellence. Futhermore, each unit or individual within a company can be viewed as providing a product or service for some other individal or unit, and the output can be evaluated using the tools of quality control.

Total quality control encompasses all eight stages of the product life cycle (Feigenbaum):

1. Marketing
2. Engineering
3. Purchasing
4. Manufacturing engineering
5. Production
6. Inspection and test
7. Shipping
8. Installation, maintenance, and service

Obviously, such a broad scope involves nearly everyone in the organization. The idea that the entire organization must be involved in the quality control effort is not new; quality experts advanced the idea as early as 1951, and the quality movement has been evolving toward TQC almost since the dawn of the industrial age. However, TQC stalled in America during the 1950s, when the quality effort was delegated to specialists in separate quality departments. Japan, on the other hand, openly embraced the TQC concept and proved that it works. This has resulted in a revival of the approach in the United States.

Quality planning is at the heart of TQC. Quality planning is an activity directed toward *preventing* quality problems. The concerns addressed by quality planning include

Establishing quality guidelines
Building quality into the design
Procurement quality
In-process and finished product quality
Inspection and test planning
Control of nonconforming material

Handling and following up on customer complaints
Education and training for quality

Establishing Quality Guidelines

Quality guidelines are established by determining customer requirements. Until the customer's wants and needs are clearly understood, it is impossible to develop a meaningful quality plan. Historically, customer needs have been determined by marketing and sales groups, often working independently of other activities within the company. Recently, however, some companies have involved production, engineering, and other groups in the effort by forming interdisciplinary teams. The joint effort has resulted in better understanding of the customer's true wants and needs and, ultimately, in better quality designs and finished goods. Once the customer's needs are known, management must determine whether company policies, procedures, and objectives are consistent with those needs. If not, they must be revised.

Building Quality into the Design

Design quality is addressed by comparing proposed design concepts with the needs of the customer, including reliability and maintainability considerations. At the same time, the design is reviewed for producibility and inspectability. It is possible to design a product that meets all of the customer's requirements but can't be made with existing technology. Sometimes the product is flawed by the use of exotic materials that are extremely costly or difficult to obtain. It is also possible to design a product that can be built, but can't be tested or inspected. If this occurs, the customer becomes the inspector, which is usually unacceptable. The design review process typically involves several departments, including marketing, manufacturing, purchasing, and quality control. Design review should be a formal activity, with mandatory sign-offs and documentation.

Procurement Quality Function

Since most buyer-seller arrangements involve contracts, the procurement quality activity tends to be quite formal. However, the importance of the less legalistic aspects of the relationship should not be overlooked. A loyal supplier, committed to your success, is something you can't get with a contract. The procurement quality function is so important that Chapter 2 of this book is devoted entirely to the subject.

In-Process and Finished Product Quality Planning

This activity involves establishing specifications for all important quality characteristics and developing formal product standards. Work instructions and detailed operating procedures are also part of this effort.

Inspection and Test Planning

Inspection and test planning is carefully integrated with the design and production activities. The work involves determination of inspection stations, classification of characteristics according to their criticality, design and procurement of inspection and test equipment, and development of inspection instructions and test procedures. These activities are typically performed by quality control or test specialists.

Control of Nonconforming Material

Nonconforming materials must be carefully segregated and identified to prevent their inadvertent mixing with acceptable materials. The material control procedures are often discussed in the purchase order or contract. Typically, a material review board composed of manufacturing, engineering, quality control, and, perhaps, a customer representative periodically reviews nonconforming materials and makes the final decision on their disposition (e.g., sort, scrap, rework). It is important to recognize that the nonconforming material is a *symptom* of some problem, and corrective action necessary to prevent future nonconforming material must be an essential element of the nonconforming material control system.

Customer Complaints

Data from customer complaints must be taken very seriously. In fact, customer complaints should be actively solicited. A major auto company discoved that a dissatisfied customer tends to complain to 14 people about his or her experience. A highly satisfied customer, on the other hand, tells only 7 others. The importance from a marketing point of view is obvious. Customer complaints are also a valuable source of information for future product improvements. Customer complaints involve two separate elements: correcting the customer's immediate problem and utilizing the data for long-term improvements.

Education and Training for Quality

W. Edwards Deming once said that most people know everything they need to know about their jobs, except how to do them better. Training in the tools of quality control, especially statistical analysis, is essential to the quality improvement effort. Quality is also affected by the amount of on-the-job training people receive, as well as by special training in job-related skills. Training is an ongoing activity that should begin before an employee is hired and continue throughout the employee's term of service.

INSPECTION AND TESTING

Inspection and testing activities have a direct impact on the quality of the product delivered to the end user. While it may be true that "you can't inspect quality into the product," you can inspect bad quality *out* of the product. Inspection and test are so important to quality that at one time they *were* quality control.

Inspection and test activity is usually directed toward one of the following goals:

Determination of product acceptance relative to customer requirements
Determination of product reliability
Qualification of a vendor, process, machine, etc.
Verification that some requirement has been met

Inspection and test involve cost. With most products, the cost of inspecting every characteristic of every unit is prohibitive. One way of dealing with this is to classify characteristics of the product and devote the bulk of attention to the characteristics most important in meeting the customer's requirements. Typically, three classifications are used: critical, major, and minor. A critical characteristic is one that has a direct impact on the safe use of the product. A major characteristic is one that is not critical but is likely to affect the usability or performance of the product. A minor characteristic is one unlikely to affect the performance of the product.

Another way to deal with the issue of economics is to sample rather than inspect 100% of the units produced. Sampling involves making decisions based on the results of inspections or tests on less than every unit. The benefit from sampling is obvious: lower costs. The opposite side of the coin is that you must tolerate a level of risk, known as sampling risk. There are two kinds of sampling risk: good products may be classified incorrectly as nonconforming (producer's risk), or nonconforming products may be classified incorrectly as conforming (consumer's risk). When dealing with critical characteristics there is no acceptable level of consumer risk, and therefore you should *never* sample critical characteristics except to verify a previous 100% inspection. An exception to this arises when the inspection or test is destructive, in which case you have no option but to sample. Once you decide to sample, you must face the question of how much to sample. This subject will be discussed in detail in subsequent chapters.

All inspection and test activity involves some degree of error. No measurement instrument is perfect, and human beings are fond of quoting the phrase " to err is human" for good reason. Interpretation of inspection and test results should be tempered by this knowledge. There are statistical methods of evaluating all sorts of inspection and test errors, but many of them are beyond the scope of this book. However, certain types of errors can easily be evaluated and controlled. For example, the accuracy of a micrometer can be checked by measuring a part of known size and adjusting the micrometer until it gives the correct result. The "part of known size" is called a standard, and the adjustment of the gage is called calibration. These subjects, and others associated with measurement errors, are discussed in Chapter 11.

Certain types of tests are classified as either destructive or nondestructive. While any test that does not do damage to the product or raw material is technically nondestructive, in practice the term *nondestructive test* is reserved for certain test procedures. The test methods most commonly classified as nondestructive are eddy current, dye penetrant, magnetic particle, ultrasonic, and X-ray tests. These tests are most often used to check properties of materials or products such as cracks and porosity.

QUALITY AND THE LAW

The subject of quality and the law is also known as product liability. We will provide only a brief overview of the subject. It is so important that Volume 2 of the *What Every Engineer Should Know* series is devoted entirely to the subject. You are encouraged to consult that book for a more comprehensive treatment of product liability. Table 1.1, reproduced from Volume 2, defines the legal terminology important in quality control.

Three legal theories are involved in product liability: breach of warranty, strict liability in tort, and negligence. Two branches of law deal with these areas, contract law and tort law. We will review the ways in which these legal concepts affect quality control.

A *contract* is a binding agreement for whose breach the law provides a remedy. In quality control work, as in product liability in general, the contract is related to the sale of a product. Contract laws for such cases are usually governed by the Uniform Commerial Code (UCC). Key concepts of contract law related to product liability are those of *breach of warranty* and *privity of contract.*

Breach of warranty can occur from either an express warranty or an implied warranty. An express warranty is part of the basis for a sale. In other words, the buyer agreed to the purchase on the reasonable assumption that the product would perform in the manner described by the seller. The seller's statement need not be written for the warranty to be an express warranty; a mere statement of fact is sufficient. An implied warranty is a warranty not stated by the seller but implied by law. Certain warranties result from the simple fact that a sale has been made. One of the most important of the attributes guaranteed by an implied warranty is that of fitness for normal use. The warranty is that the product is reasonably safe.

Privity of contract means that a direct relationship exists between two parties, typically buyer and seller. At one time, manufacturers were held not liable for products they purchased from vendors or sold to a consumer through a chain of wholesalers, dealers, etc. Manufactures were treated as third part assignees and said to be not in privity with the end user. In contract law, privity denotes parties in mutual legal relationship to one another by virtue of being promisees and promisors. This concept began to deteriorate in 1905, when courts began to permit lawsuits against sellers of unwholesome food, whether or not they were negligent, and against original manufacturers, whether or not they were in privity with the consumers. The first recognition of strict liability for an express warranty without regard to privity was enunciated by a Washington court in 1932 in a case involving a Ford Motor Company express warranty that their windshields were "shatterproof." When a windshield shattered and injured a consumer, the court

Table 1.1 Fundamental Legal Terminology

Assumption of risk: The legal theory that a person who is aware of a danger and its extent and knowingly exposes himself to it assumes all risks and cannot recover damages, even though he is injured through no fault of his own.

Contributory negligence: Negligence of the plaintiff that contributes to his injury and at common law ordinarily bars him from recovery from the defendant although the defendant may have been more negligent than the plaintiff.

Deposition: The testimony of a witness taken out of court before a person authorized to administer oaths.

Discovery: Procedures for ascertaining facts prior to the time of trial in order to eliminate the element of surprise in litigation.

Duty of care: The legal duty of every person to exercise due care for the safety of others and to avoid injury to others whenever possible.

Express warranty: A statement by a manufacturer or seller, either in writing or orally, that his product is suitable for a specific use and will perform in a specific way.

Foreseeability: The legal theory that a person may be held liable for actions that result in injury or damage only when he was able to foresee dangers and risks that could reasonably be anticipated.

Great care: The high degree of care that a very prudent and cautious person would undertake for the safety of others. Airlines, railroads, and buses typically must exercise great care.

Implied warranty: An automatic warranty, implied by law, that a manufacturer's or dealer's product is suitable for either ordinary or specific purposes and is reasonably safe for use.

Liability: An obligation to rectify or recompense for any injury or damage for which the liable person has been held responsible or for failure of a product to meet a warranty.

Negligence: Failure to exercise a reasonable amount of care or to carry out a legal duty which results in injury or property damage to another.

Obvious peril: The legal theory that a manufacturer is not required to warn prospective users of products whose use involves an obvious peril, especially those that are well known to the general public and that generally cannot be designed out of the product.

Prima facie: Such evidence as by itself would establish the claim or defense of the party if the evidence were believed.

Table 1.1 (*Continued*)

Privity: A direct contractual relationship between a seller and a buyer. If A manufactures a product that is sold to dealer B, who sells it to consumer C, privity exists between A and B and between B and C, but not between A and C.
Proximate cause: The act that is the natural and reasonably foreseeable cause of the harm or event that occurs and injures the plaintiff.
Reasonable care: The degree of care exercised by a prudent person in observance of his legal duties toward others.
Res ipsa loquitur: The permissible inference that the defendant was negligent in that "the thing speaks for itself" when the circumstances are such that ordinarily the plaintiff could not have been injured had the defendant not been at fault.
Standard of reasonable prudence: The legal theory that a person who owes a legal duty must exercise the same care that a reasonably prudent person would observe under similar circumstances.
Strict liability in tort: The legal theory that a manufacturer of a product is liable for injuries due to product defects, without the necessity of showing negligence of the manufacturer.
Subrogation: The right of a party secondarily liable to stand in the place of the creditor after he has made payment to the creditor and to enforce the creditor's right against the party primarily liable in order to obtain indemnity from him.
Tort: A wrongful act or failure to exercise due care, from which a civil legal action may result.

allowed the suit against Ford, ruling that even without privity the manufacturer was responsible for the misrepresentation, even if the misrepresentation was made innocently.

Under the rule of *strict liablity*, an innocent consumer who knows nothing about disclaimers and the requirement of giving notice to a manufacturer with whom he did not deal cannot be prevented from suing. The rule avoids the technical limits of privity, which can create a chain of lawsuits back to the party that originally put the defective product into the stream of commerce. The seller (whether a salesperson or manufacturer) is liable even though he has been careful in handling the product and even if the consumer did not deal directly with him.

This modern rule was first applied in the case of *Greenman v. Yuba Power Products, Inc.* in California in 1963. A party, Mr.Greenman, was injured when a workpiece flew from a combination power tool purchased for him by his wife two years prior to the injury. He sued the manufacturer and produced witnesses to prove that the machine was designed with inadequate setscrews.

The manufacturer, which had advertised the power tool as having "rugged construction" and "positive locks that hold through rough or precision work," claimed that is should not have to pay money damages because the plaintiff had not given it notice of breach of warranty within reasonable time as required. Futhermore, a long line of California cases held that a plaintiff could not sue someone not in privity with him unless the defective product was food.

The court replied that this was not a warranty case but a *strict liability case.* The decision stated that any "manufacturer is strictly liable . . . when an article he placed on the market, knowing that it is to be used without inspection for defects, proves to have a defect that causes injury to a human being."

The concept of strict liability was a turning point for both the consumer movement and quality control. The use of effective, modern quality control methods became a matter of paramount importance. The concept is also called *strict liability in tort,* which is virtually synonymous with the common usage of the term "product liability." A *tort* is a wrongful act or failure to exercise due care resulting in an injury, from which civil legal action may result. Tort law seeks to provide compensation to people who suffer loss because of the dangerous or unreasonable actions of others.

A related concept is that of *negligence.* Negligence occurs when one person fails to fulfill a duty owed to another or fails to act with due care. Two elements are necessary to establish negligence: a standard of care recognized by law, and a breach of the duty or requisite care. Also, the breach of duty must be the proximate cause of the harm or injury. The accepted standard of care is that of the "reasonable person." The court must measure the action of the parties involved relative to the actions expected from an imaginary reasonable person.To muddy the waters futher, the court must weigh the risk or danger of the situation against the concept of "reasonable risk." Clearly, these concepts are far from cut and dried.

The case cited above and many other developments since it have resulted in a feature that is unique to product liability law, namely: *the conduct of the manufacturer is irrelevant.* The plaintiff in a product liability suit need not prove that the manufacturer failed to exercise due care; he need show only that the product was the proximate cause of harm and that it was either defective or unreasonably dangerous. This is what is meant by strict liability. In a sense, it is the product that is on trial and not the manufacturer. The concept is defined formally as follows:

1. One who sells any product in a defective condition unreasonably dangerous to the user or consumer or to his property is subject to liability for physical harm thereby caused to the ultimate user or consumer, or to his property, if
   a. The seller is engaged in the business of selling such a product, and
   b. It is expected to and does reach the user or consumer without substantial change in the condition is which it is sold.
2. The rule in subsection (1) applies although
   a. The seller has exercised all possible care in the preparation and sale of the product, and
   b. The seller of consumer has not bought the product from or entered into any contractual relation with the seller.

As discussed in Volume 2 of this series, there are several areas in which engineering and management are vulnerable:

1. Design
2. Manufacturing and materials
3. Packaging, installation, and application
4. Warnings and labels

Designs that create hidden dangers to the user, designs that fail to comply with accepted standards, designs that exclude necessary safety features or devices, or designs that do not properly allow for possible unsafe *misuse or abuse* that is reasonably foreseeable to the designer are all suspect. Quality control includes design review as one of its major elements, and all designs should be carefully evaluated for these shortcomings. As always, the concept of reasonableness applies in all its ambiguity.

The application of quality control principles to manufacturing, materials, packaging, and shipping is probably the best protection possible against future litigation. Defect prevention is the primary objective of quality control, and the defect that isn't made will never result in loss or injury. Bear in mind, however, that a defect in quality control is usually defined as a nonconformance to requirements. There is no such definition in the law. Legal definitions of a defect are based on the concept of reasonableness and the need to consider the use of the product.

Quality control normally does not concern itself with warnings and labels. However, I have seen situations where required labels were missing or misplaced, labels didn't adhere to the surfaces to which they were affixed, and labels could not easily be seen by the end user. Such situations *are* in the domain of quality control.

SUMMARY

This chapter has introduced a working definition of quality and explained its importance. The concept of control was explained as it is related to quality control. Total quality control was introduced as an operating philosophy and guide for management. The importance of quality planning and product liability issues were discussed.

Several important topics were not covered in this chapter. The reader may wish to explore the different ways in which the basic concepts manifest themselves in different industries. Also of interest are the different applications of the basic principles in commercial firms and defense contractors. The relationship of the quality basics to important related areas such as safety, productivity, and employee relations is also worth further study.

RECOMMENDED READING LIST

30, 32–33, 40.

2

Vendor Quality Assurance

The major part of the cost of most manufactured products is in purchased materials. In some cases the percentage is as high as 80%, and it is seldom less than 50%. The importance of consistently high levels of quality in purchased materials is clear. This chapter examines important aspects of vendor quality control systems.

ETHICS AND HUMAN RELATIONS

It is important to remember that dealings between companies are really dealing between *people*. People work better together if certain ground rules are understood and followed. Above all, the behavior of both buyer and seller should reflect honesty and integrity. This is especially important in quality control, where many decisions are "judgment calls." There are certain guidelines that foster a relationship based on mutual trust:

1. Don't be too legalistic. While it is true that nearly all buyer-seller arrangements involve a contract, it is also true that unforeseen con-

ditions sometimes require that special actions be taken. If buyer and seller treat each other with respect, these situations will present no real problem.

2. Maintain open channels of communication. This involves both formal and informal channels. Formal communication includes such things as joint review of contracts and purchase order requirments by both seller and buyer teams, on-site seller and buyer visits and surveys, corrective action request and follow-up procedures, and record-keeping requirements. Informal communications involve direct contact between individuals in each company on an ongoing and routine basis. Informal communication to clarify important details, ask questions, gather background to aid in decision making, and so on, will prevent many problems.

3. The buyer should furnish the seller with detailed product descriptions. This includes drawings, workmanship standards, special processing instructions, and any other information the seller needs to provide a product of acceptable quality. The buyer should ascertain that the seller understands the requirements.

4. The buyer should objectively evaluate the seller's quality performance. This evaluation should be done in an open manner, with the full knowledge and consent of the seller. The buyer should also keep the seller informed of his *relative standing* with respect to other suppliers of the same product. However, this should be done in a manner that does not compromise the position of any other seller.

5. The buyer should be prepared to offer technical assistance to the seller, and vice versa. Such assistance may consist of on-site visits by buyer or seller teams, telephone assistance, or transfer of documents. Of course, both parties are obligated to protect the trade secrets and proprietary information they obtain from one another.

6. The seller should inform the buyer of any known departure from historical or required levels of quality.

7. The buyer should inform the seller of any change in requirements in a timely fashion.

8. The seller should be rewarded for exceptional performance. Such rewards can range from plaques to increased levels of business.

The basic principles of ethical behavior have been very nicely summarized in the Code of Ethics for members of the American Society for Quality Control, reproduced in Figure 2.1.

The American Society for Quality Control Code of Ethics

To uphold and advance the honor and dignity of the profession, and in keeping with high standards of ethical conduct ***I acknowledge that I:***

Fundamental Principles

I. Will be honest and impartial, and will serve with devotion my employer, my clients and the public.

II. Will strive to increase the competence and prestige of the profession.

III. Will use my knowledge and skill for the advancement of human welfare, and in promoting the safety and reliability of products for public use.

IV. Will earnestly endeavor to aid the work of the Society.

Relations With The Public

1.1 Will do whatever I can to promote the reliability and safety of all products that come within my jurisdiction.

1.2 Will endeavor to extend public knowledge of the work of the Society and its members that relates to the public welfare.

1.3 Will be dignified and modest in explaining my work and merit.

1.4 Will preface any public statements that I may issue by clearly indicating on whose behalf they are made.

Relations with Employers and Clients

2.1 Will act in professional matters as a faithful agent or trustee for each employer or client.

2.2 Will inform each client or employer of any business connections, interests or affiliations which might influence my judgment or impair the equitable character of my services.

2.3 Will indicate to my employer or client the adverse consequences to be expected if my professional judgment is overruled.

2.4 Will not disclose information concerning the business affairs or technical processes of any present or former employer or client without his consent.

2.5 Will not accept compensation from more than one party for the same service without the consent of all parties. If employed, I will engage in supplementary employment of consulting practice only with the consent of my employer.

Relations With Peers

3.1 Will take care that credit for the work of others is given to those to whom it is due.

3.2 Will endeavor to aid the professional development and advancement of those in my employ or under my supervision.

3.3 Will not compete unfairly with others; will extend my friendship and confidence to all associates and those with whom I have business relations.

Figure 2.1 ASQC Code of Ethics.

SCOPE OF VENDOR QUALITY CONTROL

Most companies purchase several different types of materials. Some of the materials are just supplies, not destined for use in the product to be delivered to the customer. Traditionally, vendor quality control does not apply to these supplies. Of the items destined for the product, some are simple items that have loose tolerances and an abundant history of acceptable quality. The quality of these items will usually be controlled informally, if at all. The third category of purchased goods involves items that are vital to the quality of the end product, are complex, and have limited or no history. Purchase of these items may even involve purchase of the vendor's expertise (designs, application advice, etc). It is the quality of this category of items that will be the subject of subsequent discussions.

VENDOR QUALITY SYSTEMS

One early question is always, "who is responsible for vendor quality?" When it is asked in this way, the result is usually much pointless discussion. Instead of beginning with such a broad question, it is usually better to break down the tasks involved in assuring vendor quality and then assign responsibility for the individual tasks. The chart below is one such breakdown (Juran, 1980, p. 229).

Responsibility Matrix

Department			
Product design	Purchasing	Quality control	Activity
X	X	XX	Establish a vendor quality policy.
	XX		Use multiple vendors.
X	X	XX	Evaluate quality capability of potential vendors.
XX		X	Specify requirements for vendors.
X		XX	Conduct joint quality planning.
	X	XX	Conduct vendor surveillance.
X		XX	Evaluate delivered product.
X	X	XX	Conduct improvement programs.
	XX	X	Use vendor quality ratings in selecting vendors.

X = shared responsibility.
XX = primary responsibility.

It is important to recognize that, in the end, the responsibility for quality always remains with the supplier. The existence of vendor quality systems for "assuring vendor quality" in no way absolves the vendor of responsibility. Rather, these systems should be viewed as an aid to the vendor.

MULTIPLE VENDORS

The subject of whether or not to use multiple vendors is one that arouses strong feelings both pro and con. For decades, the conventional wisdon in American quality control was that multiple vendors would keep all suppliers "on their toes" through competition. Also, having multiple vendors was looked at as a hedge against unforeseen problems like fire, flood, or labor disputes. These beliefs were so strong that multiple vendors became the de facto standard for most major government agencies, including the Department of Defense.

In the 1980s the consensus on multiple sources of supply began to erode. Japan's enormous success with manufacturing in general and quality in particular inspired American businesspeople to study the Japanese way of doing things. Among the things Japanese businesses do differently is that they *discourage* multiple source purchases whenever possible. This is in keeping with the philosophy of W. Edwards Deming, a noted American quality expert who provided consulting and training to many Japanese firms. The advocates of single-source procurement argue that it encourages the supplier to take long-term actions on your behalf and makes suppliers more loyal and committed to your success. A statistical argument can also be made: minimum variability in product can be obtained if the sources of variation are minimized, and different suppliers are an obvious source of variation.

Since both sides have obvious strong points, the decision regarding single-source versus multiple-source procurement is one that must be made on a case-by-case basis. Your company will have to examine its unique set of circumstances in light of the arguments on both sides. In most cases a policy of using single sources except under unusual conditions works well.

EVALUATING VENDOR QUALITY CAPABILITY

When making an important purchase, most companies want some sort of advance assurance that things will work out well. When it comes to quality, the vendor quality survey is the "crystal ball" used to provide this assurance. The vendor quality survey usually involves a visit to the vendor by a team from the buyer prior to the award of a contract; for this reason it is sometimes called *preaward survey*. The team is usually composed of representatives from the buyer's design engineering, quality control, production, and purchasing departments. The quality control elements of the survey usually include, at a minimum,

Quality management
Design and configuration control
Incoming material control
Manufacturing and process control
Inspection and test procedures
Control of nonconforming material
Gage calibration and control
Quality information systems and records
Corrective action systems

The evaluation typically involves use of a checklist and some numerical rating scheme. A simplified example of a supplier evalutation checklist is shown in Figure 2.2.

The checklist in Figure 2.2 is very simple compared to those used in practice. Many checklists run on to 15 pages or more. My personal experience is that these checklists are very cumbersome and difficult to use. If you are not bound by some government or contract requirement, I recommend that you prepare a brief checklist similar to the one shown and supplement it with a narrative that documents your personal observations. Properly used, the checklist can help guide you without tying your hands.

Bear in mind that a checklist can never substitute for the knowledge of a skilled and experienced evaluator. Numerical scores should be supplemented by the observations and interpretaions of the evaluator. The input of vendor personnel should also be included. If there is disagreement between the evaluator and the vendor, the positions of both sides should be clearly described.

In spite of their tremendous popularity, physical vendor surveys are only one means of evaluating the potential performance of a supplier. Studies of their accuracy suggest that they should be taken with a large grain of salt. One such study by Brainard (1966) showed that 74 of 151 vendor surveys resulted in incorrect predictions (i.e, either a good supplier was predicted to be bad or vice versa)—a coin flip would have been as good a predictor! I recommend that physical vendor surveys be avoided whenever possible. One option is to use so-called desk surveys. With desk surveys you have the prospective suppliers themselves submit lists of test equipment, certifications, organization charts, manufacturing procedures, etc. Check these documents over for obvious problems.

Vendor Evaluation Checklist	
criteria	result
The quality system is developed	
The quality system is implemented	
Personnel are able to identify problems	
Personnel recommend and initiate solutions	
Effective quality plans exist	
Inspection stations are identified	
Management regularly reviews quality program status	
Contracts are reviewed for special quality requirements	
Processes are adequately documented	
Documentation is reviewed by quality	
Quality records accurate and up to date	
Effective corrective action system	
Non-conforming material properly controlled	
Quality costs properly reported	
Changes to requirements controlled	
Adequate gage calibration control	

Figure 2.2 Vendor Evaluation Checklist. Since this type of evaluation is conducted at the supplier's facility, it is known as a physical survey.

Regardless of the survey method used or the outcome of the survey, it is important to keep close tabs on the actual quality history. Have the vendor submit "correlation samples" with the first shipments. The samples should be numbered and each should be accompanied by a document showing the vendor's quality inspection results and test results. Verify that the vendor has correctly checked each important characteristic and that your results agree, or "correlate." Finally, keep a running history of quality performance (see Quality Records later in this chapter). The best predictor of future good performance seems to be a record of good performance in the past. If you are a subscriber to the Government, Industry Data Exchange Program (GIDEP) you have access to a wealth of data on many suppliers. (GIDEP subscribers must also contribute to the data bank.) Another data bank is the Coordinated Aerospace Supplier Evaluation (CASE). If relevant to your application, these compilations of the experience and audits of a large number of companies can be a real money-and time-saver.

VENDOR QUALITY PLANNING

Vendor quality planning involves efforts directed toward preventing quality problems, appraisal of product at the vendor's plant as well as at the buyer's place of business, corrective action, dispostition of nonconforming merchandise, and quality improvement. The process usually begins in earnest after a particular source has been selected Most preaward evalutation is general in nature; after the vendor is selected it is time to get down to the detailed level.

A first step in the process is the transmission of the buyer's requirements to the vendor.Even if the preliminary appraisal of the vendor's capability indicated that the vendor could meet your requirements, it is important that the requirements be studied in detail again before actual work begins. Close contact is required between the buyer and the vendor to assure that the requirements are clearly understood. The vendor's input should be solicited; could a change in requirements help the vendor produce better quality parts?

Next it is necessary to work with the vendor to establish procedures for inspection, test, and acceptance of the product. How will the product be inspected? What workmanship standards are to be applied? What in-process testing and inspection is required? What level of sampling will be employed? These and similar questions must be answered at this stage. It is good practice to have the first few parts completely inspected by both the vendor and the buyer to assure that the vendor knows which features must be checked as well as how to check them. The buyer many want to be at the vendor's facility when production first begins.

Corrective action systems must also be developed. Many companies have their own forms, procedures, etc. for corrective action. If you want your system to be used in lieu of the vendor's corrective action system, the vendor must be notified. Bear in mind that the vendor many need addtional training to use your system. Also, the vendor may want some type of compensation for changing his established way of doing things. If at all possible, let the vendor use his own systems.

At an early stage, the vendor should be made aware of any program you have that would enable him to reduce inspection. At times it is possible to certify the vendor so that *no* inspection is required by the buyer and shipments can go directly to stores, bypassing receiving inspection completely.

If special record keeping will be required, this must be spelled out. Many companies have special requirements imposed on them by the nature of their business. For example, most major defense items have traceability and configuration control requirements. Government agencies such as the Food and Drug Administration (FDA) often have special requirements. Automotive companies have record-keeping requirements designed to facilitate possible future recalls. The vendor is often in the dark regarding your special record-keeping requirements; it is your job to keep him informed.

POSTAWARD SURVEILLANCE

The focus up to this point has been on developing a process that will minimize the probability of the vendor producing items that don't meet your requirements. This effort must continue after the vendor begins production. However, after production has begun the emphasis can shift from an evaluation of systems to an evaluation of *actual* program, process, and product performance.

Program evaluation is the study of a supplier's facilities, personnel, and quality systems. While this is the major thrust during the preaward phase of an evaluation, program evaluation doesn't end when the contract is awarded. Change is inevitable, and the buyer should be kept informed of changes in the vendor's program. Typically, this is accomplished by providing the buyer with a registered copy of the vendor's quality manual, which is updated routinely (see Chapter 4 for futher information on quality manuals). Periodic follow-up audits may also be required, especially if product quality indicates a failure of the quality program.

A second type of surveillance involves surveillance of the vendor's process. Process evaluations involve a study of methods used to produce an end result. Process performance can usually be best evaluated by statistical methods, and it is becoming common to require that statistical process control (SPC) be applied to critical process characteristics (see Chapter 7). Many large companies require that their suppliers perform statistical process control studies, called process capability studies, as part of the preaward evaluation.

The final evaluation, product evaluation, is also the most important. Product evaluation consists of evaluating conformance to requirements. This may involve inspection at the vendor's site, submission of objective evidence of conformance by the vendor, inspection at the buyer's receiving dock, or actual use of the product by the buyer or the end user. This *must* be the final proof of performance. If the end product falls short of requirements, it matters little that the vendor's program look good or that all of the in-process testing meets established requirements.

Bear in mind that the surveillance activity is a communications tool. To be effective it must be conducted in an ethical manner, with the full knowledge and cooperation of the vendor. The usual business communication techniques, such as advanced notification of visits, management presentations, exit briefings, and follow-up reports, should be utilized to assure complete understanding.

SPECIAL PROCESSES

We will define a special process as one that has an effect that can't readily be determined by inspection or testing subsequent to processing. The difficulty may be due to some physical constraint, such as the difficulty in verifying grain size in a heat-treated metal, or the problem may simply be economics, such as the cost of 100% X-ray of every weld in an assembly. In these cases special precautions are required to assure that processing is carried out in accordance with requirements.

The two most common approaches to control of special processes are certification and process audit. Certification can be applied to the skills of key personnel, such as welder certification, or to the processes themselves. With processes the certification is usually based on some demonstrated capability of the process to perform a specified task. For example, a lathe may machine a special test part designed to simulate product characteristics otherwise difficult or impossible to measure. The vendor is usually responsible for certification. Process audit involves establishing a procedure for the special process, then reviewing actual process performance for compliance to the procedure. A number of books exist to help with the evalutation of special processes. In addition, there are inspection service companies from which you can hire experts to verify that special processes meet established guidelines. These companies employ retired quality control professionals as well as full-time personnel. In addition to reducing your costs, these companies can provide a level of expertise you may not otherwise have.

QUALITY RECORDS

Reference was made earlier in this chapter to the importance of tracking actual quality performance. In fact. this "track record" is the most important basis for evaluating a vendor and the best predictor of future quality performance. Such an important task should be given careful consideration. The purpose of vendor quality records is to provide a basis for action. The data may indicate a need to reverse an undesirable trend or may suggest a cost savings from reduced (or no) inspection. Also, the records can provide an objective means of evaluating the effectiveness of corrective action.

An early step is to determine what data will be required. Data considerations include analyzing the level of detail required to accomplish the objectives of the data collection described above (i.e., how much detail is needed to determine an appropriate course of action?). Some details normally included are part number, vendor identification, lot identification number, name of characteristic(s) being monitored, quantity shipped, quantity inspected, number defective in the sample, lot disposition (accept or reject), and action taken on rejected lots (sort, scrap, rework, return to vendor, use as is). These data are usually first recorded on a paper form and later entered into a computerized data base. The original paper documents are kept on file for some period of time, the length of time being dictated by contract, warranty, tax, and product liability considerations.

The data collection burden can be significant, especially in larger companies. Some companies have resorted to modern technology to decrease the work load. For example, common defects may have bar codes entered on menus for the inspector. The vendor will apply a bar code sticker with important information such as the purchase order number and lot number, which the computer can use to access a data base for additional details on the order. Thoughtful use of a bar code system can reduce the data input chore to a few keystrokes. Some bar code systems have been designed for use on personal computers, bringing the technology within the reach of nearly everyone.

Although the raw data will contain all of the information necessary to provide a basis for action, the information will be in a form that lacks meaning to the decision makers. To make the raw data meaningful, they must be analyzed and reported in a way that makes the information content obvious. Reports should be designed with the end reader in mind. The data reported should be carefully considered and only the necessary information should be provided. In most cases it will take several reports to adequately distribute the information the right people. Typically, the level of detail provided will decrease as the level of management increases. Thus, senior management will usually be most interest in summary trends, perhaps presented as graphs and charts rather than numbers. Operating personnel will receive reports that provide information on specific parts and characteristics. At the lowest level, audit personnel will receive reports that allow them to verify the accuracy of the data entered into the system.

The general subject of data collection and reporting is covered in great detail in Chapter 5.

VENDOR RATING SCHEMES

As the discussion so far makes clear, evaluating vendors involves comparing a large number of factors, some quantitative and some qualitative. Vendor rating schemes attempt to simplify this task by condensing the most important factors into a single number, the vendor rating, that can be used to evaluate the performance of a single vendor over time or to compare multiple sources of the same item.

Most vendor rating systems involve assigning weights to different important measures of performance, such as quality, cost, and delivery. The weights are selected to reflect the relative importance of each measure. Once the weights are determined, the performance measure is multiplied by the weight and the results totaled to get the rating. For example, we might decide to set up the following rating scheme:

Performance	Measure	Weight
Quality	% of lots accepted	5
Cost	Lowest cost/cost	300
Delivery	% on-time shipments	2

The calculations for three hypothetical suppliers are shown below.

Vendor	% lots accepted	Price	% on-time deliveries
A	90	$60	80
B	100	70	100
C	85	50	95

Vendor	Quality Rating	Price Rating	Delivery Rating	Total Rating
A	450	250	160	860
B	500	214	200	914
C	425	300	190	915

As you can see, the simple example above combines reject rates, delivery performance, and dollars into a single composite number. Using this value, we would conclude that vendors B and C are approximately the same. What vendor B lacks in the pricing category is made up for in quality and delivery. Vendor a has a much lower rating than B or C.

A valid question at this point would be, "is all of this discussion meaningful?" Obviously, one of the problems is the mixture of units. As one quality control text puts it, "Apples + Oranges = Fruit salad." Other questions involve the selection of weights, disposition of rejected lots (scrap, rework, use as is, return to vendor, etc.), number of lots involved, cost of defects, types of defects, how late the deliveries were, whether early deliveries and late deliveries both count against the delivery score, and so forth. It is easier to answer such qustions if some guidelines are provided. Here are some characteristics of good rating schemes:

The scheme is clearly defined and understood by both the buyer and the seller.
Only relevant information is included.
The plan is easy to use and update.
The scheme is applied only where it is needed.
The ratings "make sense" when viewed in light of other known facts.

Finally, I must express some personal skepticism regarding rating schemes. In my 20 years of work in the quality profession, I have never seen a rating scheme that didn't have many "holes" in it that required subjective interpretation which defeats the purpose of the rating. Also, there are very few cases where the ratings can be usefully applied: multiple sources of the same important item over a time period sufficient to provide a valid basis for comparison. Many ratings have values that are highly questionable; for example, a best-selling book on quality control shows a model rating scheme that is based in part on "percent of promises kept." Ratings are usually easy for vendors to manipulate by varying lot sizes or delivery dates or selectively sorting lots. Finally, and I believe most damning, rating schemes tend to draw attention from the true objective, never-ending improvement in quality.

PROCUREMENT STANDARDS AND SPECIFICATIONS

Quality control of purchased materials is an area that has been well explored. As a result, many of the basics of supplier quality control have been drafted into standards and specifications that describe many of the details common to most buyer and seller transactions. We will discuss some examples of commercial specifications as well as government specifications.

Most large companies have formalized at least some of their quality control requirements for purchased material. Traditionally, their specifications have covered the generic requirements of their supplier's quality system, including the details of any given purchase in the purchase order or contract. These supplier quality specifications typically cover such things as the seller's quality policy, quality planning, inspection and test requirements, quality manual, audits, the *supplier's* material control system, special process control, defective material control, design change control, and corrective action systems. The scope of the different specifications varies considerably from company to company, as does the level of detail. A good representative example is ANSI/ASQC Z1.15, Generic Guidelines for Quality Systems, available from the American Society for Quality Control, 310 W Wisconsin Avenue, Milwaukee, WI 53203.

The Department of Defense has formalized its quality system requirements in several different documents. The best known is Mil-Q-9858, which describes the general requirements for quality systems for prime contractors. Another standard, Mil-I-45208, describes the requirements for inspection systems. Mil-I-45208 is a subset of Mil-Q-9858.

SUMMARY

This chapter explained the importance of personal integrity in vendor quality assurance. The scope of vendor quality control was discussed. The separate and joint responsibilities of vendor and vendee were outlined. Some methods of evaluating vendor quality were introduced and their strengths and weaknesses described. Vendor rating systems were also covered. A few standards and specifications for procurement activities were mentioned.

This chapter did not describe detailed implementation of vendor quality control programs. Also omitted were the specific contents of procurement standards and specifications. The important topic of vendor quality audit was introduced but not covered in depth.

RECOMMENDED READING LIST

1-5, 26-29.

3

Human Resources and Quality

Most engineers in every field discover the human element very early in their careers. It often seems that a major part of the engineer's job is to "engineer the human being out of the system." Of course, no matter how hard we try, the ubiquitous human element usually remains. In fact, there are important functions performed *best* by people!

The situation is no different with quality control; human beings are a major source of quality problems. However, just as in the design activity, the quality system can be engineered to minimize the problems caused by human errors. Also as in the design activity, there are many important quality-related activities that humans still perform better than any machine. This chapter discusses the role of people in quality control and quality improvement.

BASIC PRINCIPLES OF INDUSTRIAL PSYCHOLOGY

There are several theories of human behavior vying for recognition. Although the science of psychology is over 200 years old, the effort to understand and explain human behavior is still in its infancy. Still, it has much to offer anyone interested in motivating people to do a better job.

Maslow's Hierarchy of Needs

Professor A. S. Maslow of Brandeis University has developed a theory of human motivation that has been elaborated on by Douglas McGregor. The theory describes a "hierarchy of needs." Figure 3.1 illustrates this concept.

Maslow postulates that the lower needs must be satisfied before one can be motivated at higher levels. Futhermore, as an individual moves up the hierarchy the motivational strategy must be modified because *a satisfied need is no longer a motivator.*

The hierarchy begins with physiological needs. At this level a person is seeking the simple physical necessitites of life, such as food, shelter, and clothing. A person whose basic physiological needs are unmet will not be motivated with appeals to personal pride. If you wish to motivate personnel at this level, provide monetary rewards such as bonuses for good quality. Other motivations include opportunities for additional work, promotions, or simple pay increases. As firms con-

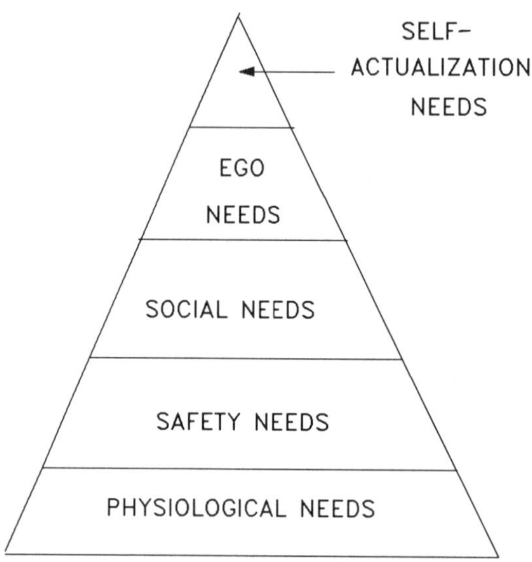

Figure 3.1 Hierarchy of Needs.

tinue to do more business in underdeveloped regions of the world, this category of worker will become more commonplace.

Once the simple physiological needs have been met, motivation tends to be based on safety. At this stage issues such as job security become important. Quality motivation of workers in this stage was once difficult. However, since the loss of millions of jobs to foreign competitors who offer better quality goods, it is easy for people to see the relationship between quality, sales, and jobs.

Social needs involve the need to consider oneself as an accepted member of a group. People who are at this level of the hierarchy will respond to group situations and work well on quality circles, employee involvement groups, or quality improvement teams.

The next level, ego needs, comprise needs for self-respect and the respect of others. People at this level are motivated by their own craftsmanship as well as by recognition of their achievements by others.

The highest level is that of self-actualization. People at this level are self-motivated. This type of person is characterized by creative self-expression. All you need do to "motivate" someone at this level is to provide an opportunity for him or her to make a contribution.

Herzberg's Hygiene Theory

Frederick Herzberg is generally given credit for a theory of motivation known as the hygiene theory. The underlying assumption of the hygiene theory is that job satisfaction and job dissatisfaction are not opposites. Satisfaction can be increased by paying attention to "satisfiers," and dissatisfaction can be reduced by dealing with "dissatisfiers." The theory is illustrated in Figure 3.2.

Theories X, Y, and Z

All people seem to seek a coherent set of beliefs that explain the world they see. The belief systems of managers were classified by McGregor into two categories, which he called Theory X and Theory Y.

Under Theory X workers have no interest in work in general, including the quality of their work. Because civilization has mitigated the challenges of nature, modern humans have become lazy and soft. The job of managers is to deal with this by using "carrots and sticks." The carrot is monetary incentive, such as piece rate pay. The stick is docked pay for poor quality or missed production targets. Only money can motivate the lazy, disinterested worker.

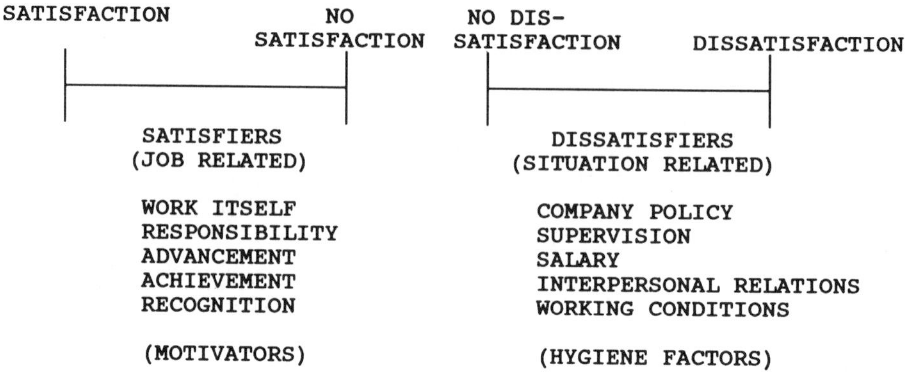

Figure 3.2 Herzberg's hygiene theory.

Theory Y advocates believe that workers are internally motivated. They take satisfaction in their work and would like to perform at their best. Symptoms of indifference are a result of the modern workplace, which restricts what workers can do and separates them from the final results of their efforts. It is management's job to change the workplace so that the workers can, once again, recapture their pride of workmanship.

Theories X and Y have been around for decades. Much later, in the 1980s, Theory Z came into vogue. The Z organizations have consistent cultures where relationships are holistic, egalitarian, and based on trust. Since the goals of the organization are obvious to everyone and integrated into each person's belief system, self-direction is predominant. In the Z organization, Theories X and Y become irrelevant. Workers don't need the direction of Theory X management, nor does management need to work on removal of barriers, since there are none.

OPERATOR- VERSUS MANAGEMENT-CONTROLLABLE PROBLEMS

All the theories notwithstanding, a basic question that must be answered early is whether or not a particular class of problem is something we should expect the operator to be able to avoid. An operator-controllable problem has three distinct traits (Juran and Gryna, 1980):

1. The operators know what they are supposed to do.
2. The operators know what they are actually doing.
3. The operators have the responsibility, authority, skill, and tools necessary to correct the problems.

If any one of these traits is absent, then the problem is a management-controllable problem. Let's examine each of these three traits in more detail.

The Operators Know What They Are Supposed to Do

How often have you heard someone say, "if they would just do their jobs, everything would work out fine!" The number of lawsuits for breach of contract attests to the difficulty in clearly defining the requirements for any given assignment. Learning what the job is can be a difficult task.

Before operators can know what they are supposed to be doing, *management* must determine what the operators are supposed to be doing. This involves developing detailed procedures for the tasks that need to be performed and then training the operators in the proper interpretation of the procedures. The procedures must be written down; simply telling an operator what to do when he starts the job is not enough. It is also good practice to test the operator's understanding of the instructions. The test may be written, but verbal tests and demonstrations by actually doing the task under observation are also used.

Another aspect of knowing what is supposed to be done is the inspection standard, or the operational standard. Just how are "good" and "bad" defined? The definition should be as clear and unambiguous as possible, a task that is never as easy as it looks. For example, how would you define the printing standard for this book? You may find that such phrases as "the letters must be dark and crisp" mean different things to different people. Before an operator can know what he is supposed to do, such ambiguities must be dealt with. Until operator, inspector, supervisor, management, and end user all agree on the operational standard, the problem is a management problem, not an operator-controllable problem.

No company or institution operates solely by written procedures. Much of what is done involves *operating precedents*. An operating precedent is an established way of doing things. In all companies the operating precedent carries tremendous weight. If the written instructions contradict the operating precedents, the operator is placed in a very awkward position. Following the operating precedent means violating the written procedure, and following the written procedure means violating the operating precedent. For example, the written procedure might say to run a line at 5 to 7 units per hour, but it is traditional to run the line 10 units per hour at the end of the month when schedules get tight, even though the error rate tends to increase at the higher rate. If the operator has authority to change the line speed, he can be blamed for the additional errors if he sets the line to 10 units, but he'll also be the scapegoat for missed deliveries if he insists on holding to the written limits of 5 to 7 units. Because the resolution of such inconsistencies is beyond the control of the operator, the presence of any inconsistency between written practice and operating precedent make the problem a management problem, not an operator problem.

The Operators Know What They Are Actually Doing

Feedback is the essence of control. Unless the process output is evaluated, at least periodically, there is no basis for corrective action. Accurate timely feedback must be available to the operator if the problem is to be classified as an operator problem.

There are two basic types of process feedback data: variables measurements and attributes. Variables measurements are numbers like length, width, temperature, and weight. Attributes are characteristics possessed (or not possessed) by the process or the product. For example, a product may have a scratch, paint may be blistered, or a cleaning process may pass or fail a cleanliness test such as a water break test (if the water beads up the cleaning process fails, if the water sheets off the process passes).

If the operators are to use variables measurements to determine how they are doing, they must be given information on how many units to sample, how often to sample, what measurements to take, how to record the measurements, and perhaps what statistical process control analysis to perform on the data. Gaging or measurment systems must be provided and must be accurate and repeatable enough to provide a good basis for process control (see Chapter 11 for addtional details on evaluating measurement systems). The operators should be properly trained in the use of the measurement systems. If attribute data are to be taken, many of the same questions must still be answered; the operators must still know how many samples to check, how often to check the process, what type of statistical process control (SPC) analysis to use, etc.

Operators should get more than the immediate real-time feedback they need for process control. Trend reports and other long-term analyses should be made available to the operators. The analysis necessary for these reports is usually performed by the operators' supervisor or the quality control department, not the operators themselves. Also, if any additonal information is generated downstream, such as at the next operation or from the customer's experience with the output from the process, the operators should receive this information too. In other words, management should do all it can to assure that the operator gets complete information to use in process control and process improvement.

The Operators Have the Responsibility, Authority, Skill, and Tools Necessary to Correct the Problems

This means that when a problem occurs, the operators know what action is required to correct it and have the ability and authority to take the corrective action. Futhermore, if a problem occurs that an operator *can't* fix, he has the authority to take an alternative action, such as shutting down the process.

When action is taken, the operator must also have some feedback on the effectiveness of the action. This implies that the process reacts in a measurable way to the action. In most cases the feedback should come from the measurements in the "operators know what they are doing" section above. However, in some cases there must be faster feedback. For example, a manufacturer of ceramic parts has an operator quality control system. If the part quality indicates, the operator must adjust the temperature of a firing furnace. However, the temperature change must occur in a special zone within the furnace, and the change can be verified only with a special instrument normally kept in a laboratory. Since it takes several hours for a load of parts to run through the furnace, provisions are made for the operator to call for a special temperature check if he makes a temperature change.

Even though it is popular to blame the operator for most problems, the reader should know that most quality problems are *not* operator controllable. Approximately 80% to 85% of all quality problems can be solved only by management action. By applying the criteria described above, you will be able to separate problems into each category and thus assign responsibility properly. Chapter 7 discusses additional statistical criteria for making this determination.

CATEGORIES OF HUMAN ERRORS

When trying to eliminate or reduce errors, it is often helpful to "divide and conquer." By carefully dividing errors into different categories, we can sometimes better see what type of action is appropriate.

Inadvertent Errors

Many human errors occur because of lack of attention. People are notorious for their propensity to commit this type of error. Inadvertent errors have certain hallmarks:

> There is usually no advance knowledge that an error is imminent.
> The incidence of error is relatively small. That is, the task in normally performed without error.
> The occurrence of errors is random, in a statistical sense.

Examples of inadvertent errors are not hard to find. This is the type of error we all make in everyday life when we find a mistake in balancing our checkbook, miss a turn on a frequently traveled route, dial a wrong number on the phone, or forget to pay a bill. These things can be overlooked at home, but not in quality control.

Preventing inadvertent errors may seem an impossible task. Indeed, these errors are among the most difficult of all to eliminate. And the closer the error rate gets to zero, the more difficult improvement becomes. Still, in most cases it is possible to make substantial improvements economically. At times it is even possible to eliminate the errors completely.

One way of dealing with inadvertent errors is *foolproofing*. Foolproofing involves changing the design of a process or product to make the commission of a particular human error impossible. For example, a company was experiencing a sporadic problem (note: the words "sporadic problem" should raise a flag in your mind that inadvertent human error is likely!) with circuit board defects. It seems that occasionally entire orders were lost because the circuit boards were drilled wrong. A study revealed that the problem occurred because the circuit boards could be mounted backward on an automatic drill unless the manufacturing procedure was followed carefully. Most of the time there was no problem, but as people became more experienced with the drills they sometimes got careless. The problem was solved by adding an extra hole in an unused area of the circuit board panel and then adding a pin to the drill fixture. If the board was mounted wrong, the pin wouldn't go through the hole. Result: no more orders lost.

Another method of reducing human errors is automation. People tend to commit more errors when working on dull, repetitive tasks. Also, people tend to make more errors when working in unpleasant environments, where the unpleasantness may arise from heat, odors, noise, or a variety of other factors. It happens that machines are very well suited to exactly this type of work. As time passes, more progress is being made with robotics. Robots redefine the word "repetitive." A highly complicated task for a normal machine becomes a simple repetitive task for a robot. Elimination of errors is one item in the justification of an investment in robots. (Volume 11 of this series is devoted to robots). On a more mundane level, using simpler types of automation such as numerically controlled machining centers often produces a reduction in human errors.

Another approach to the human error problem is ergonomics, or human factors engineering. Many errors can be prevented through the application of engineering principles to design of products, processes, and workplaces. By evaluating such things as seating, lighting, and sound levels, temperature change, and workstation layout, the environment can often be improved and errors reduced. Sometimes human factors engineering can be combined with automation to reduce errors. This involves automatic inspection and the use of alarms (lights, buzzers, etc.) that warn the operator when he's made an error. This approach is often considerably less expensive than full automation.

Technique Errors

I was once involved with a problem with gearbox housing. The housings were gray iron castings and the problem was cracks. Our supplier was made aware of the problem, and the supplier's metallurgist and engineering staff worked long and hard on the problem, but to no avail. Finally, in desperation, we sat down with the supplier to put together a "last-gasp" plan. If our plan failed, we simpy had to try an alternative source for the casting.

As you'd expect, the plan was grand. We identified many important variables in the product, process, and raw materials. Each variable was classified as either a "control variable", which we would take steps to hold constant, or an "experimental variable," which we would vary in a prescribed way. The results of the experiment were to be analyzed using all the muscle of a major mainframe statistical analysis package. All of the members of the team were confident that no stone had been left unturned.

Shortly after the program began, I received a call from our quality engineering representative at the supplier's foundry. "We can continue with the experiment if you really want to," he said, "but I think we've identified the problem and it isn't on our list of variables." It seems that the engineer was in the inspection room inspecting castings for our project and he noticed a loud "clanging sound" in the next room. The clanging occurred only a few times each day, but the engineer soon noticed that the cracked castings came shortly after the clanging began. Finally, he investigated and found that the clanging sound was a relief operator pounding the casting with a hammer to remove the sand core. Sure enough, the cracked castings has all received the "hammer treatment"!

This example illustrates a category of human error different from the inadvertent errors described earlier. Technique errors share certain common features:

They are unintentional.
They are usually confined to a single characteristic (e.g., cracks) or class of characteristics.
They are often isolated to a few workers who consistently fail.

Solution of technique errors involves the same basic approaches as the solution of inadvertent errors, namely automation, foolproofing, and human factors engineering. In the meantime, unlike inadvertent errors, technique errors may be caused by a simple lack of understanding that can be corrected by developing better instructions and training.

Willful Errors (Sabotage)

This category of error is unlike either of the two previous categories. Willful errors are often very difficult to detect; however, they do bear certain trademarks:

They are not random.
They "don't make sense" from an engineering point of view.
They are difficult to detect.
Usually only a single worker is involved.
They begin at once.
They do not occur when an observer is present.

Perhaps an example would be helpful. An electromechanical assembly suddenly began to fail on some farm equipment. Examination of the failures revealed that the wire had been broken *inside the insulation*. However, the assemblies were checked 100% after the wire was installed and the open circuit should have been easily discovered. After a long and difficult investigation, no solution was found. However, the problem went away and never came back.

About a year later I was at a company party when a worker approached me. He said he knew the answer to the "broken wire mystery," as it had come to be known. The problem was caused, he said, when a newly hired probationary employee was given his two weeks notice. The employee decided to get even by sabotaging the product. He did this by carefully breaking the wires, but not the insulation, and then pushing the broken sections together so that the assembly would pass the test. However, in the field the break would eventually separate, resulting in failure. Later, I checked the manufacturing dates and found that every failed assembly had been made during the two weeks prior to our saboteur's termination date.

Often, the security specialist is far better equipped and trained to deal with this type of error than quality control or engineering personnel. In serious cases, criminal charges may be brought as a result of sabotage. If the product is being made on a government contract, federal agencies may be called in. Fortunately, willful errors are extremely rare. They should be considered a possibility only after all other explanations have been investigated and ruled out.

SUMMARY

This chapter described the basic principles of industrial psychology that are especially relevant to quality control. The basic theories of management behavior, Theories X, Y, and Z, were discussed. Features that distinguish operator- and management-controllable problems from one another were described. A simple method of classifying and preventing human errors was presented.

The chapter did not discuss cultural and anthropological concerns, which may be very important for firms doing business in foreign countries or in disadvantaged areas of the United States. Also not covered were the myriad specific approaches to employee motivation such as employee involvement or quality circles. Human factors engineering methods, while introduced, were not described in detail.

RECOMMENDED READING LIST

15-16, 34-39.

4

Quality Organization and Management

This chapter describes the basic tasks required to assure that products and services meet the quality objectives of management. The emphasis is necessarily on tasks. However this presents a danger that must be understood very clearly: *Quality must be judged in terms of the final result, not in terms of completed tasks!*

The quality of your product or service, as judged by the end user, is the final criterion for measuring success. This simple lesson is absolutely critical to the long-term viability of the company. It is extremely easy for a company to inadvertently slip into the trap of judging quality by asking if everyone has met their responsibilities. That misses the point of the quality effort, namely to achieve quality judged by the consumer to be worth the cost relative to that of your competitors.

QUALITY POLICY

The first step in the quality management process is up to top management. What is the company's policy toward quality? While this may seem simple, it is the essence of the entire quality effort. The Policy should be given a great deal of thought. Platitudes won't do ("XYZ corporation believes in quality"). The quality policy should be a guide to everyone in the company. If the policy is to provide competitive quality at a lower price, then the decisions of marketing, engineering, and production will be different from the decisions that would be made if the policy was to be the quality leader at a premium price. Everything from the advertising campaign to vendor selection to employee training depends on the policy of top management.

Policy statements are usually quite vague. This need not be the case. Although the company policy statement is not the place for detailed dissussions of responsiblity, it must be specific enough to provide the required guidance to everyone in the company. Ford Motor Company's mission statement is a case in point:

> Ford Motor Company is a worldwide leader in automotive and automotive related products and services, as well as in newer industries such as aerospace, communications, and financial services. Our mission is to improve continually our products and services to meet our customer's needs allowing us to prosper as a business and to provide a reasonable return for our stockholders, the owners of our business.

A number of things are made clear by this mission statement. First, the focus of the company is external. The position of leadership can only be measured in relation to the company's competitors. The customer's needs guide the company. The company's scope is defined. And Ford plans to provide a reasonable return consistent with its mission. This mission statement tells employees that the status quo will not be tolerated as a performance standard. They must seek ways to improve their products and services relative to the competition, and they must evaluate their success by looking to the customer. All of this must be done while providing a reasonable return on the investment of the stockholder, which is not necessarily the highest in the industry.

Perhaps the best way to illustrate the guidance provided by this particular mission statement is to contrast it with another one. Let's create a mythical company, "Econo Car Corporation." The policy of this company might state:

> Econo Car Corporation is a world leader in providing transportation to the cost-conscious consumer. Our mission is to provide safe, basic transportation at the lowest price.

The difference in basic philosophy is obvious. What is less apparent is the impact of this on day-to-day decisions. Nearly every action taken by every employee would be influenced to some degree by the policy. Everything from the paper used by the office typing pool to the target audience for the advertising would be affected. Clearly, the company policy statement should be given a great deal of thought. There should be nothing in the policy statement that is not endorsed by *every member* of the senior staff. Consensus is an absolute requirement.

COMPANY ORGANIZATION FOR QUALITY

In the preindustrial age the craftsman sold directly to the end user, whom he usually knew personally. With this arrangement there was no need for any kind of formal organization, for quality or any other reason. This began to change with the industrial revolution. Manufacture of interchangeable parts required at least a rudimentary inspection activity to assure that the various parts fit together. Factories were built in such a way that different activities were performed in different areas, such as the foundry, the machine shop, and assembly. Organizations tended to mimic the layout of the factory. In the early 1900s mass production methods were pioneered by Henry Ford and the management systems adjusted accordingly. "Scientific management" emerged as the dominant management style. Among other things, scientific management involves the division of labor and organization along the lines of the tasks managed.

Early organizaitons reflected the minor role of quality. Quality, when it was considered at all, was deemed to be the responsibility of the production department. When scientific management caught on, people with the title of "inspector" began to appear. An inspector was often the last person on the assembly line. The problem with this arrangement was that the inspector reported to the person who was responsible for production. Often, merchandise that was unfit for use was shipped anyhow.

The next step in this evolution was to create a separate department for inspection. The inspection department was usually headed by an inspection foreman who reported to the plant manager. In larger organizations, a chief inspector position was created. The chief inspector supervised the activities of several inspection foreman. The inspection-oriented quality organization persisted between World War I and World War II.

The existence of an independent inspection activity highlighted the cost of rejects. The obvious waste made many people ask if the defects couldn't be prevented from occurring in the first place. World War II dramatized the waste. Furthermore, techniques developed by Walter A. Shewhart of Bell Laboratories provided a scientific means of preventing defects through process control. Shewhart's methods of statistical process control, supplemented by other statistical methods developed during the 1930s and 1940, became known collectively as statistical quality control, or SQC, methods. During the postwar period there was tremendous enthusiasm for SQC. The emphasis was clearly different from that of inspection. Also, the people who filled the positions were, typically, engineers. The job title of many of these new professionals was "quality control engineer." The organization structure was duly changed to accommodate the new positions in a separate department known as the quality engineering department. Quality engineering, along with the inspection department, reported to a new position, quality control manager.

The complexity of products and systems continued to increase following World War II. Many of the new systems failed in service from problems not related to the traditional arena of quality control, which was manufacturing. The failures were inherent in the design. It became necessary to broaden the scope of quality-related activities to include the design function. Quality planning became more widespread, and a new specialty, reliability engineering, made its appearance. Once again, the organization changed to accommodate these new activities.

The next major change involved increasing activism by the consumer. As consumers gained political strength, legislation was passed making effective quality control a requirement rather than an option. Regulatory bodies imposed standard requirements for quality control, such as the FDA's Good Manufacturing Practice (GMP). Other agencies were created to act as advocates for consumers, such as the Consumer Product Safety Commission. Agencies became much more aggressive on behalf of their consumer constituency and recalls of defective products became commonplace, costing manufacturers billions of dollars. Finally, changes in tort law made it easier for consumers to bring (and win) suits against manufacturers for placing defective products into the stream of commerce. All of these things acted to push the visibility of quality upward on the organization chart. By 1980 it was common to have vice-presidents of quality in larger organizations, and even quite small organizations had quality managers reporting directly to the chief executive officer. A typical organization chart is shown in Figure 4.1.

Work Elements

The work elements of quality control are divided among the organizational units in Figure 4.1 as shown in Table 4.1.

TASKS AND THE TOTAL QUALITY SYSTEM

The organization chart and discussion above are misleading in that they suggest that quality is the responsibility of specialists in the quality department. In fact, this is decidely *not* the case. As Chapter 1 emphasized, the specialists in the quality department have no more than a secondary responsibility for most of the important tasks that affect quality. The role of others in the company can be better understood if we look at a listing of the basic tasks as part of a system, which we will refer to as the total quality system (TQS). The TQS can be viewed as a process that assures continual improvement while implementing the policy established by top management. The TQS process is depicted in Figure 4.2.

The TQS makes it clear that the quality department shares the responsibility for quality with many others throughout the company. Let's examine Figure 4.2 more closely.

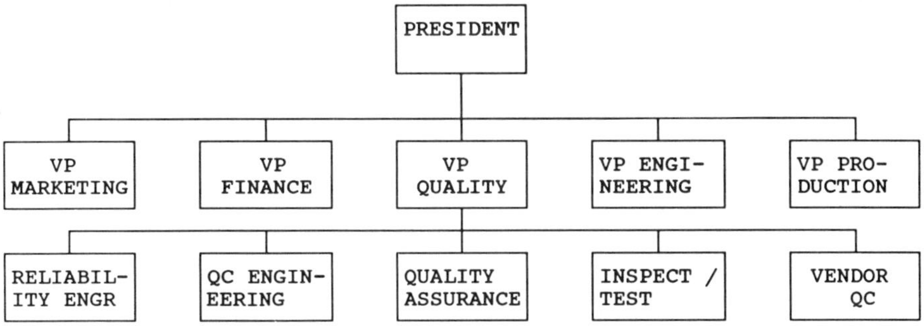

Figure 4.1 Quality Organization Chart

Table 4.1 Work Elements

Reliability engineering work elements
Establishing reliability goals
Reliability apportionment
Stress analysis
Identification of critical parts
Failure mode, effects, and criticality analysis (FMECA)
Reliability prediction
Design review
Supplier selection
Control of reliability during manufacturing
Reliability testing
Failure reporting and corrective action system
QC engineering work elements
Process capability analysis
Quality planning
Establishing quality standards
Test equipment and gage design
Quality troubleshooting
Analysis of rejected or returned material
Special studies (measurement error, etc.)
Quality assurance work elements
Write quality procedures
Maintain quality manual
Perform quality audits
Quality information systems
Quality certification
Training
Quality cost systems
Inspection and test work elements
In-process inspection and test
Final product inspection and test
Receiving inspection
Maintenance of inspection records
Gage calibration
Vendor quality control
Pre-award vendor surveys
Vendor quality information systems
Vendor surveillance
Source inspection

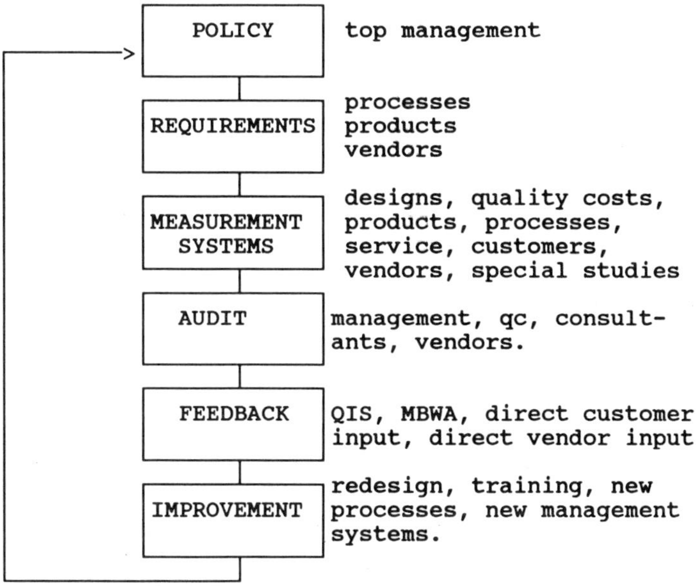

QIS = Quality Information Systems.
MBWA = Management By Wandering Around.

Figure 4.2 Total Quality System (TQS)

TQS Policy

As discussed earlier, top management is responsible for establishing the company policy on quality. This policy forms the basis of the requirements.

TQS Requirements

Requirements are established for key products, processes, and vendors. Product requirements are based on market research regarding the customer's needs and the ways in which competitors' products meet or fail to meet these needs. Depending on the policy of the company, a product design must either compete with existing products or improve on them. Process requirements assure that designs are reproduced with a minimum of variability, where "minimum" is interpreted in a manner consistent with established policy. Process design is usually the responsibility of process engineering specialits. Vendor selection is typically based on reliability, price, and quality. Purchasing bears the primary responsiblity, with quality control and engineering providing assistance.

TQS Measurement Systems

The term "measurement system" is used here in the broadest sense. The purpose of a measurement system is to determine the degree to which a product, service, vendor, process, or management system conforms to the requirements established. At times the measurement is not straightforward. For example, design conformance is accomplished by a design review. Quality control, marketing, and manufacturing evaluate the design to determine whether it meets the customer's expectations and can be economically and consistently reproduced by manufacturing. This is a subjective evaluation, but the parties involved should reach a consensus.

Quality cost measurement is such an important subject that it is discussed in depth later in this chapter. Suffice it to say that the design of quality cost systems is usually the joint responsibility of the accounting and quality control departments.

Product and process measurements are designed to assure that the products and processes conform to requirements. This usually involves obtaining one or more observations and comparing the observations to operational standards. Although the quality control department may be responsible for obtaining measurements on certain products and processes, the responsibility for acting on the results remains with the operating department.

Measurement of service quality has been woefully inadequate. In fact, many companies fail to recongize their responsibility to the customer in this area. In most cases of neglect it takes an external force, such as a government agency or a competitor, to create a change. Unlike the situation in most of the other areas, responsibility for measuring service quality has not been established by precedent. However, it seems logical that quality control and marketing share a joint responsibility.

Measurement of vendor quality was discussed in Chapter 2.

Quality control and purchasing were the primary players in developing the measurement systems. Of course, the final responsibility for quality remained with the vendors themselves.

Special studies are different from the other measurement systems described here in that they are not an ongoing activity. Studies may be commissioned to answer an important question or to obtain vital facts. For example, a competitior may have just released a revolutionary product that requires close examination. The responsibility for special studies is usually stipulated in the charter that commissions the study.

Finally, remember that as stated early in this chapter, the proper emphasis is on results, not task completion. The most important measurement is therefore, customer satisfaction. Marketing is usually given responsibility for developing systems that measure customer satisfaction.

TQS Audit

The purpose of a TQS audit is to seek verification that an adequate system exists and is being adhered to. The primary responsibility for auditing TQS is top management's. In addition, the manager of each activity must perform audits within his or her area to assure that the requirements are being met. Finally, "third party" audits by the quality department or auditors from outside the company are also common. All third party audits should be commissioned by management. However, the third party audit should never be the primary source of detailed information for management. There is no substitute for firsthand observation.

Basic ground rules must be established before proceeding with an audit. First, there must actually be a requirement against which performance is audited. The performance standard should be written and communicated to the responsible parties. Second, there must be an established means of measuring performance. Finally, the responsibility for action must be unambiguous and clearly understood. Third party audits should be preannounced whenever possible.

Audits can be divided into two basic categories. The *quantitative audit* is designed to provide a score that indicates the degree of compliance of a program or quality system. The *qualitative audit* is a generalized assessment of the effectiveness of a quality program. Since in practice a quality program has elements that are both quantitative and qualitative, it is rare to find any audit that is purely qualitative or quantitative; most are a mixture of both. In general, the qualitative audit is less formal, and it is the type of audit that should be conducted by internal personnel, including management, auditing their own operations. The quantitiative audit is normally conducted by third parties, and it provides an objective assessment that can be used to spot trends in performance.

Audits serve many different purposes. At a financial level they can provide a means of evaluating the effectiveness of quality cost expenditures. Properly done, audits can provide a means of motivating people to higher levels of performance. Audit results can provide a guide to action by identifying system elements that might be the cause of problems. Audits can be used as an early warning system that indicates potential trouble before it actually leads to a negative customer reaction.

Audits can also be classified according to just what is being audited. At the broadest level, the *quality program audit* assesses the appropriateness of the quality program in relation to the company policy and mission. Quality programs range from minimal to extremely elaborate, and it may be that the formal quality program does not fit the company's objectives or industry. Given that the quality program accurately reflects the wishes of top management regarding scope and size, the *quality system audit* can be conducted. This audit measures the overall effectiveness of the quality program and provides guidance for improving the effectiveness. *Audits of procedures* are conducted to determine the extent of compliance to procedures. At times such audits uncover procedures that are outdated or ineffective or show a need for formal procedures where none currently exist. Also common are *product audits*. Product audits can take place either while the product is in process or on the finished product. Product audits are usually in-depth evaluations of very small samples to determine the extent of compliance to established product requirements. The product auditor should perform the evaluation from the viewpoint of the end user or, at times, of the outside regulator. *Process audits* are conducted to evaluate the conformance of materials and processes to established requirements. Statistical process control (SPC) by operators is a type of process audit, as is the periodic review of control charts by area managers. Finally, *special audits* are often commissioned by management to get needed facts on the company's operations. These audits can range from opinion surveys to housekeeping checks, or any other subject that interests management.

TQS Feedback

All of the information collection in the world is absolutely useless without an effective feedback mechanism. The feedback loop of the total quality system is closed by a variety of means. The most visible is the formal quality information system (QIS) of the company. The QIS encompasses all of the formal, routine activities associated with quality data collection, analysis, and dissemination. The subject is a big one and is covered in detail in Chapter 5.

Formal systems, while necessary, provide only part of the feedback. Human beings may get the facts from these systems, but they can't get a *feeling* for the situation from them. This level of feedback comes only from firsthand observation. One company encourages its senior managers, in fact *all* of its managers, to "wander around" the work areas to observe and to interact with the people doing the work. The term given to this activity is management by wandering around, or MBWA. I can personally attest to the effectiveness of this approach, having done it myself and seen it work. There are many "facts" that never appear in any formal report. Knowing that quality costs are 10% of sales is not the same as looking at row upon row of inspectors, or a room full of parts on hold for rework, or a railroad car of scrap pulling away from the dock.

Just as MBWA provides direct feedback from the workplace, a company needs direct personal input from its customers and suppliers. Company personnel should be encouraged to visit customers. All managers should be required periodically to answer customer complaints (after being properly trained, of course). A stint on the customer complaint "hot line" will often provide insight that can't be obtained in any other way. Vendors should be asked to visit your facility. More formal methods of getting customer and vendor input are also important. Vendors should be made part of the process from design review to postsale follow-up. Often a slight design modification can make a part much easier to produce, and the vendor, who knows his processes better than you do, will often suggest this modification if he's invited to do so. Customer input should be solicited through opinion surveys, customer review panels, and the like. Of course, the QIS will provide some of this input.

Improvement

The whole point of getting feedback is improvement. Why bother collecting, analyzing, and distributing information if it will not be used? This simple fact is often overlooked, and many companies devote vast resources to developing ever more sophisticated management information systems that produce little or no improvement. Discovering this situation is one of the primary objectives of the quality audits discussed earlier.

Continuous improvement should be a routine part of everyone's job, but making this actually happen is extremely difficult. Most operating precedents and formal procedures are designed to maintain the status quo. Systems are established to detect negative departures from the status quo and react to them. Continuous improvement implies that we constantly attempt to change the status quo for the better. Doing this wisely requires an understanding of the nature of cause systems. Systems will always exhibit variable levels of performance, but the nature of the variation provides the key to the type of action that is appropriate. If a system is "in control" in a statistical sense, then all of the observed variability is from common causes of variation that are inherent in the system itself. Improving performance when this situation exists calls for fundamental changes to the system. Other systems will exhibit variability that is clearly nonrandom in nature. Variation of this sort is said to be due to "special causes" of variation. When this is true it is usually best to identify the special cause rather than to take any action to change the system itself. Changing the system when variability is from special causes is usually counterproductive, as is looking for "the problem" when the variability is from common causes. Determining whether variability is from special causes or common causes requires an understanding of statistical methods. Some of the more useful and easily understood methods are described in subsequent chapters of this book.

Quality improvement must be companywide in scope. The entire cycle from marketing and design through installation and service should be geared toward improvement. This holds regardless of the policy of the company. Companies that aspire to be cost leaders have as much need for improvement within their market segment as those that aim for quality leadership. A key part of this effort must be continuous improvement of the management systems themselves.

THE QUALITY MANUAL

Formal quality systems are documented in the quality manual. The manual contains company policies and procedures that affect product quality. Quality audits by third parties often begin with the quality manual. Most quality manuals follow a similar format. For example, the table of contents shown in Table 4.2 is from the *Manual of Quality*

Table 4.2 Contents of a Quality Manual

Introduction.
Administration.
Quality Procedures and Bulletins.
Stamp and signature administration.
Audit and surveillance administration.
Training and certification administration.
Quality program management.
Organization and functions.
Quality planning.
Work instructions.
Records and reports.
Corrective action.
Costs related to quality.
Facilities and standards.
Drawings, documentation and changes.
Measuring and test equipment.
Production tooling as media of inspection.
Control of purchases.
Candidate supplier/subcontractor quality evaluation and selection.
Active supplier/subcontractor quality compliance.
Manufacturing control.
Materials and material control.
Production processing and fabrication.
Completed item inspection and testing.
Handling, storage and delivery.
Nonconforming material.
Statistical quality control and analysis.
Coordinated customer/contractor actions.
Customer inspection at subcontractor or vendor facilities.
Customer property.

Assurance Procedures and Forms (Carlson et al., 1981), which is based on the requirements of Mil-Q-9858A, a military standard for the quality programs of prime contractors. Quality manuals must be updated constantly. Typically, manuals are serialized and those with registered manuals are sent bulletins as changes are made.

In my work I have seen dozens of quality manuals. Subsequent audits of the systems documented by these manuals lead me to believe that the size of the manual has no relationship to the effectiveness of the quality system. Good quality manuals, like the effective systems they document, are well thought out and customized to the particular needs of the company. The emphasis is on clarity and common sense, not on an effort to document every conceivable eventuality. For example, the quality manual of a very large defense contractor listed *68 ways* to identify scrap material (red paint, stamped with an "R," a scrap tag, etc.). This approach led to constant difficulty because there were always new cases not allowed for in detail. In one instance a government auditor complained because some defective microchips were not properly identified. The chips were in a plastic bag, and even though a properly completed scrap tag was in the bag with them, the parts themselves were not stamped with an "R" (it's hard to stamp a part only a fraction of an inch square!). As you might expect, the company quality control manual was several hundred pages thick and very few people outside the audit department had ever read it. In contrast, I audited a medium-sized electronics company whose quality manual consisted of 45 well-written pages. The scrap identification problem was handled by a single sentence that stated: "Scrap material will be clearly and permanently identified." In my audit of the quality systems I found that all registered manual holders had read the manual and understood it.

QUALITY COSTS

The American Society for Quality Control (ASQC) divides quality costs into four categories:

Prevention costs: Costs incurred to prevent the occurrence of nonconformances in the future

Appraisal costs: Costs incurred in measuring and controlling current production to assure conformance to requirements

Internal failure costs: Costs generated before a product is shipped as a result of nonconformance to requirements

External failure costs: Costs generated after a product is shipped as a result of nonconformance to requirements

These cost categories allow the use of quality cost data for a variety of purposes. Quality costs can be used for measurement of progress, for analysis of problems, or for budgeting. By analyzing the relative size of the cost categories, the company can determine if its resources are properly allocated.

As stated earlier, the quality cost system should be developed jointly by the quality and accounting departments. Although the task of monitoring quality costs on an ongoing basis is normally the responsibility of the accounting department, in some companies the quality department takes the lead. Because of this, the quality cost system is often thought of as an accounting system or a quality department system. This is a mistake. The quality cost system is primarily a *management information system* whose purpose is to guide management in their efforts to improve quality. The point of making this distinction is that one should not go to extremes in trying to get to-the-penny accuracy. By their nature, quality costs are often difficult to determine, and at times a good approximation is the best you will be able to obtain. Usually, this is all you need to determine the appropriate management action. The practice of overemphasizing to-the-penny accuracy is a common cause of failure of quality cost programs (Juran and Gryna, 1980, pp. 30–31).

Quality cost elements, while sometimes difficult to measure, are not difficult to categorize. Table 4.3 provides a guide to some of the more commonly encountered quality cost elements.

Quality costs are an extremely important means of directing management action. They can also be very useful in helping management track the success of their quality improvement efforts. Ideally, the total cost of quality (COQ) will decline over time. Crosby (1979) recommends a 10% per year goal for reducing total COQ, but this figure seems somewhat arbitrary. A better approach would be to make a realistic assessment of the *causes* of current quality cost elements and develop an improvement plan based on this assessment. This approach requires an in-depth understanding of quality technology by upper management and the use of SPC and other statistical methods introduced in this book. One especially helpful quality tool is the quality information system (see Chapter 5), which should provide the detailed data on which the COQ figures are based.

Table 4.3 Quality Cost Elements

Prevention costs
- quality planning
- process control planning
- design review
- quality training
- gage design

Appraisal costs
- receiving inspection
- laboratory acceptance testing
- in-process inspection
- quality audits
- outside endorsements (e.g. Laboratory approval)
- calibration
- inspection and test equipment
- field testing

Internal failure costs
- scrap
- rework
- process troubleshooting
- vendor caused scrap or rework
- material review board activity
- re-inspection or re-test
- downgrading

External failure costs
- processing of customer complaints
- service
- unplanned field repair
- recalls
- processing of returned materials
- warranty

IMPACT OF QUALITY COSTS ON PROFITABILITY*

The basic factors of production are land, labor, capital, and (some economists include) management. Note that these factors also con-

*"Impact of Quality Cost Reductions on Profits." Copyright 1976, American Society for Quality Control, Inc., Milwaukee, Wisconsin. Reprinted by permission from American Society for Quality Control.

stitute the source of all cost. Thus most managers equate cost to output and assume that the two are inextricably related. If we call costs "inputs" and the production that results from the inputs "output," then the model of production in the minds of most managers is

Increased input = increased output

An analogous mental model of profitability is that it is inextricably linked to sales volume. This model can be described with the equation

Increased sales = increased profit

From these two equations it seems to follow that the most direct means available for increasing the absolute level of profits is to increase the absolute level of sales. As we shall see, in many cases this assumption is absolutely untrue.

An alternative means of increasing profit is to reduce costs. Cost reduction effected through a given program often seems trivial when expressed as a percentage of total sales. However, the direct impact of cost reduction is *not* on sales. Obviously, cost reductions have their most direct impact in the area of costs (by definition). Also note that costs are the converse of profits. The sum of costs and profits equals sales (in a poorly managed company, the sum of costs and losses).

Since the ratio of profit to sales is often large, it is important that the impact of a cost reduction be stated in a way that accurately reflects its true importance, which is, to repeat, in the area of cost reduction and profitability and *not* in the area of sales. It was stated that sales equals costs plus profits. The usual justification for an increase in costs (investment) is that it will earn a return. Generally, this means an increase in the profitabiliy figure that will someday exceed the adverse impact on costs. Usually this assumes an increase in sales. The logic is that profit is a function of sales. The common phenomenon of early loss on new investment is revealed in the sales formula restated to express the profit relationship:

Profit = sales − costs

The expected increase in sales suffers only a production lag but the increase in profits expected from a new investment always suffers a cost lag, frequently called a "payoff period."

Given this normal relationship, we can analyze the effect of a hypothetical investment on sales, costs, and profits. Let us assume that our investment will yield a return of 20% and that it will be externally funded and not taken from current profits. We assume that prior to our investment our firm's financial statement looked like this:

Sales	\$10,000
Cost	9,500
Profit	500

We further assume that the new investment will only expand the capacity of our firm and will not affect the percentage of return on sales. This means that we will continue to have profitability of 5% return on sales as shown above even after we invest in the additional capacity. Given these assumptions, here is what the financial statement would look like after we invested the \$500:

Sales	\$12,000
Cost	11,400
Profit	600

You can see that the increase in profit of \$100 constitutes a 20% return on the \$500 investment. If we are to maintain our 5% return on sales, this profitability must come from an addtional \$2000 in sales. Since cost = sales − profit, the cost of operating the new capacity must be \$11,400. This is an additional cost of \$1900, more than 2½ times the \$500 investment! It is the operations costs that offer great potential for improved profitability.

In the foregoing example we attempted to increase profit by increasing sales. Note, however, that in the example the sales figure is 20 times the profit figure. Also, the relationship of sales, profit, and cost is linear. Given this linear relationship, it is apparent that cost changes that do not affect the sales figure will alter the profit figure by an amount equal to the change in costs. Most costs have a beneficial effect on sales, but there is one category that does not: failure costs. This fact is significant in that failure cost reductions become, in effect, technological improvements, and they enable firms to overcome the adverse impact on profits caused by diminishing returns to scale.

When one input is variable, the relationship between input and product is conventionally divided into three stages, as illustrated in Figure 4.3.

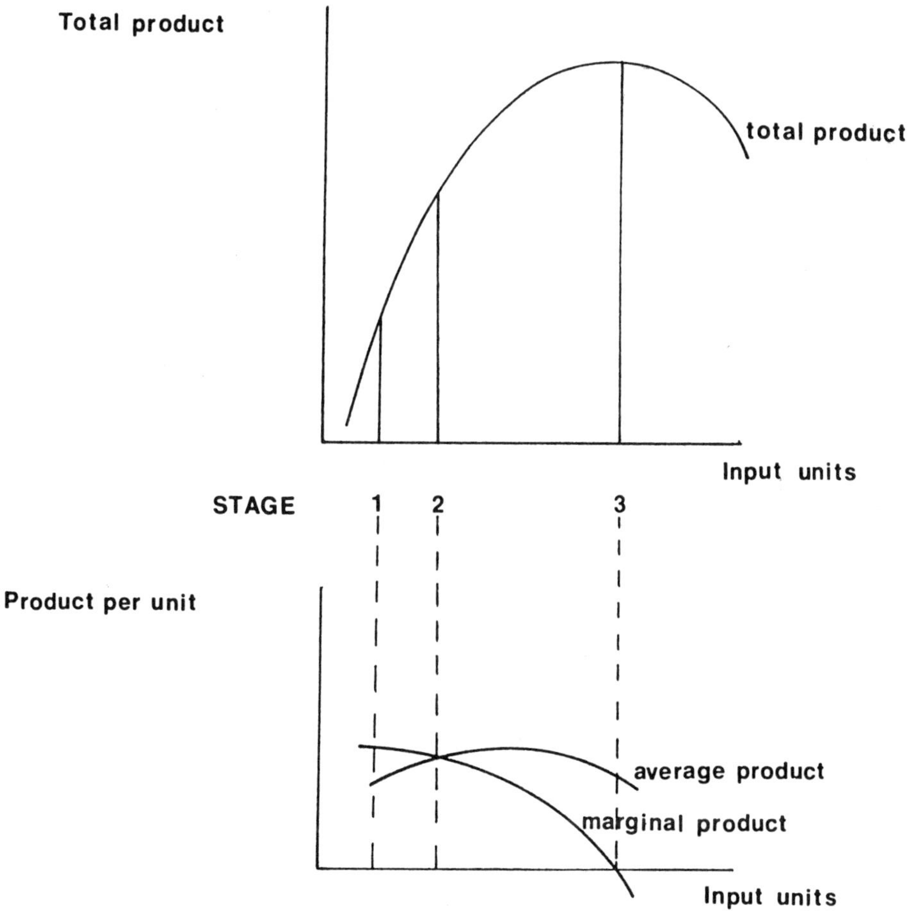

Figure 4.3 Stages in the life of a firm.

In stage 1 the average product per input increases. For example, two workers would produce more than double the output of one worker. The point where marginal and average products are equal and average product is at a maximum is the boundary of stage 1.

In stage 2, total product continues to increase but at a decreasing rate. The right-hand boundary of stage 2 is at maximum total product and zero marginal product. In stage 2 also, both average and marginal product are on the decline; marginal product, being below the average product, pulls the average down.

In stage 3, total product is declining. No rational producer operates for any period of time in stage 3. A firm that is near the boundary of stage 3 has little to gain from an increase in sales. Any sales increase will require increased output to meet it, and this will move the producer even closer to stage 3 and negative marginal product. The producer would actually suffer a *loss* from increased sales.

When a firm is in a market dominated by a few sellers (oligopoly), increased output from new capacity presents a different problem. In this case unsatisfied demand must exist prior to the increase in output since it cannot be created via price cuts. In oligopolistic markets the competitors would simply follow suit, and the result would be to *increase* the cost fugure while *decreasing* the dollar volume of sales. Thus, new investment in a market characterized by oligopoy is predicated on the assumption that demand for the product will increase given the current price.

Obviously, a beneficial impact on profit in the foregoing situations cannot be best realized by increasing output on sales. The input variable required is technology, which alters the relationship between factor inputs and marginal productivity. In looking at Figure 4.3, we see that a favorable impact on marginal productivity can move a firm *backward* on the growth curve to a position of more dynamic growth, including growth in profit.

Consider some reasons for this phenomenon. The level of profit is increased in both relative and absolute terms; more investment is attracted because of the higher returns; consumer demand that went unsatisfied because of the negative impact of diminishing returns to scale can now be met; firms improve their positions relative to their competitiors; and so on, ad infinitum.

Failure Costs

Two categories of investment produce the beneficial effect described above: R & D investment and investment in the reduction of failure costs. We are, of course, concerned with the latter category. Reduction of failure costs has a direct impact on the bottom line because *it has no adverse effect on output*. Like an R & D investment, reduction of failure costs produces new technology in the sense that more output is obtained from less input. Usually an equivalent impact from return on sales requires a much greater investment, and, as we have seen, often sales dollars are unable to make *any* profit contribution.

In economics the term *opportunity cost* is defined as the dollar amount that would be derived from the employment of a factor of production in the best alternative use. In the purest sense, failure costs are opportunity costs, lost units of output from factors of production. And the opportunity forgone is profit.

Management deserves to be informed of these opportunity costs and of their true importance. Management is used to equating profit to sales. It can be an eye-opener if, in addition to the customary measurement bases, management is told what volume of sales it would take to generate as much profit as is forgone in failure (waste) costs.

The figures show the potential improvement in profitability. Realizing this potential requires an investment in quality technology. Investing in higher profit through improved ability to eliminate waste is every bit as sensible as investing in additional capacity and sales. In many cases, it is the only sensible way to improve competitiveness.

SUMMARY

The importance of developing a consistent and meaningful quality policy was described. The evolution of the structure of quality organizations over time was covered. Work elements of the various groups in the typical quality organization were presented. Also covered were organizations for implementing the TQS process. Quality audit systems were described along with feedback systems. Quality manuals were introduced and their contents examined. Quality cost measurement systems were described, and the impact of quality costs on profitability was evaluated.

The chapter did not discuss industry-specific organizations for quality or the many different ways in which quality responsibilities are allocated in different companies.

RECOMMENDED READING LIST

1-7, 23-25, 30, 32, 40.

Quality Information Systems

A quality information system (QIS) is used in collecting, analyzing, reporting, or storing information for the total quality system. This chapter discusses the fundamental principles of QIS, followed by a case study of one company's experience in establishing a QIS from scratch.

The QIS contains both manual and computerized elements and requires inputs from many different functions within the company. In many companies the QIS is one part of an even more cmprehensive system known as the management information system or MIS. The MIS is an effort to bring the data collection activities of the entire organization into a single integrated system. When it works properly, the MIS has a number of advantages. For one thing, it eliminates the duplication of effort that would be required if the MIS did not exist. Also, it is a globally optimized system (i.e., the *whole system* is designed to be of maximum benefit to the company). Without an MIS the information system is likely to be a collection of optimized subsystems that are "best" for the individual departments they serve but are probably not optimal for the company as a whole.

Management information systems also have their darker side. Even the concept itself can be questioned. "Dictatorship of data processing" is one common complaint. In designing the one grand system, data processing has the ultimate say in whose needs are served and when they will be served. Second, MISs are often so large and complex that changing them is nearly impossible. Finally; MISs are concerned almost completely with computerized systems; this makes them incomplete, since much vital information never makes it into a computerized data base.

PLANNING A QIS

Because a QIS requires inputs from a wide variety of different functions within the company, planning the system will involve a large number of people from these different groups. The quality department should not develop a QIS in a vacuum and expect it to be accepted. The QIS is not a quality department system, it is a *companywide system*. However, the quality department usually bears the major responsibilty for operating the system.

Although QIS means "computerized" to most people it is important to note that most successful QIS applications begin as successful *manual* applications. Usually a system element important enough to be made a part of the computerized QIS is too important to wait until it can be programmed; thus some form of manual system is established at least as a stopgap measure. Planning the manual system is no trivial task; data collection needs must be established, personnel selected and trained, reports designed and their distribution lists established, and so forth. Because of this vested work, manual systems provide a great deal of information that can be used in planning the more elaborate computer-based sytem.

The flowchart is a tool that has been used in planning computer systems since the beginning of the computer age. Flowcharts can be applied at all stages in the process, from system design to writing the actual computer program code. The standard symbols used in creating information system flowcharts are established by the American National Standards Institute (ANSI) and are shown in Figure 5.1.

A general approach to QIS planning is shown in Figure 5.2. Good QIS planning always starts with the benefits of the system. This drives the rest of the process. When possible, benefits should be expressed in dollars. However, the most important benefits are sometimes the most

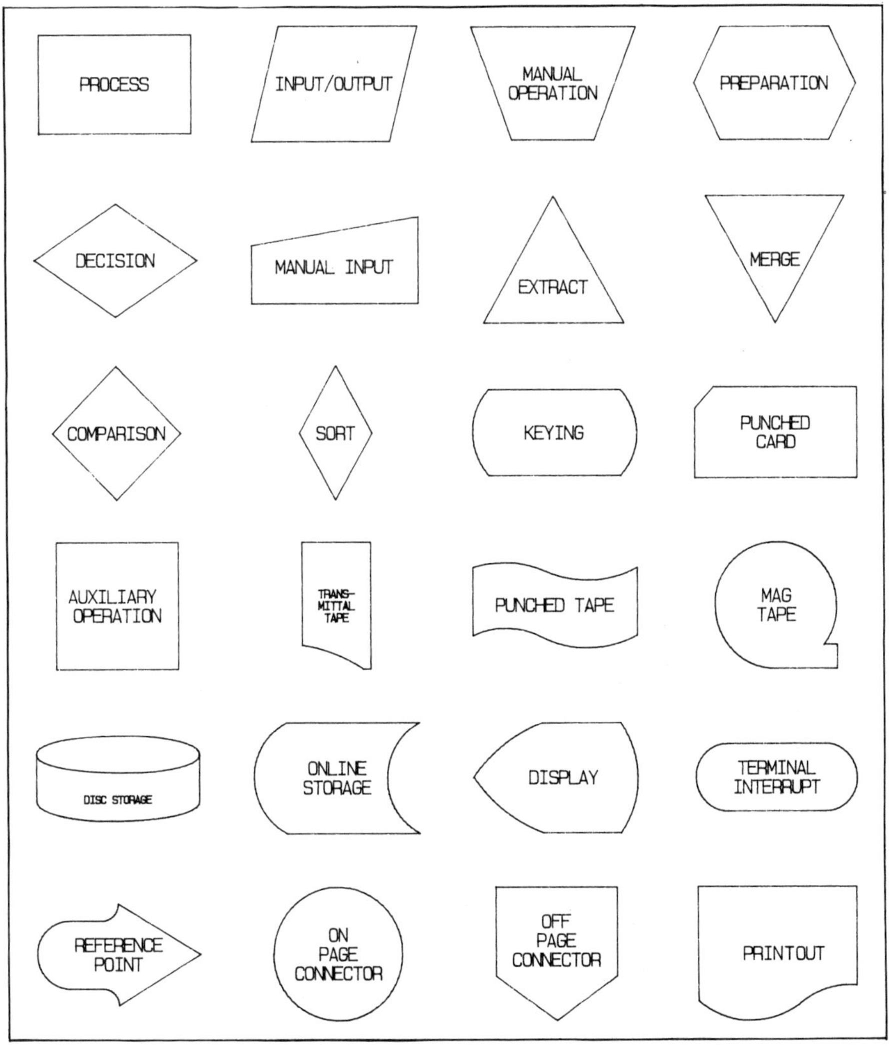

Figure 5.1 Flowchart symbols.

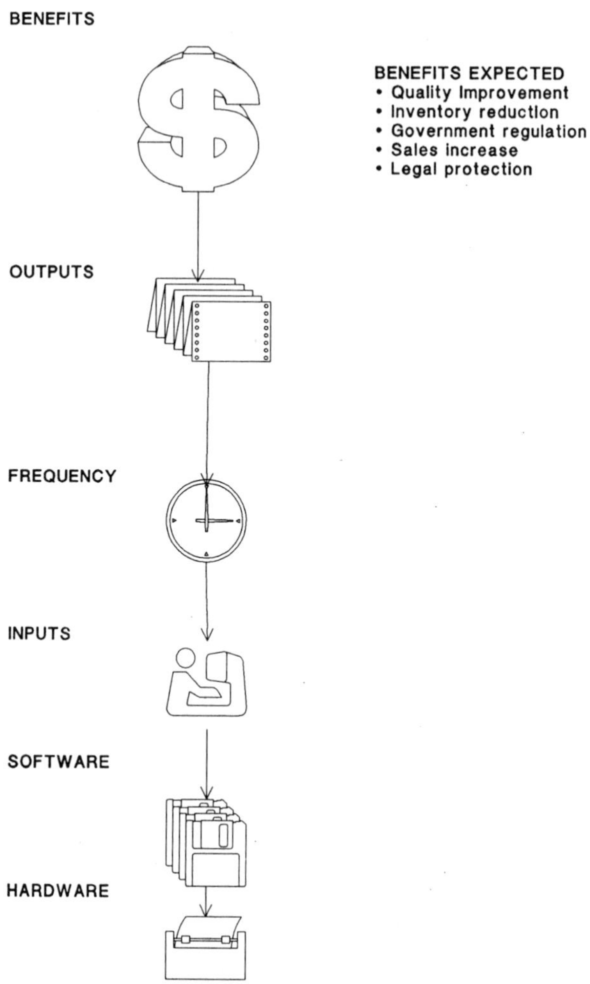

Figure 5.2 Planning flowchart for a QIS.

difficult to quantify, such as safety. Other QIS elements are required by law—for example, FDA traceability requirements.

After deciding that the benefits justify the creation of the QIS element, the next step is to define the outputs required. This should be done in as much detail as possible, as it will be useful in later stages of the planning process. Output requirements should be classified into "must have" and "would like to have" categories. Often the cost of a system is increased dramatically because of the effort required to add some optional item. Data processing professionals refer to these options as "bells and whistles." The form of the output—on-line data, periodic reports, graphics, etc.—can have a significant impact on the total cost of the system.

How often the output needed is also important. Continuous output in the form of a real-time system is at one extreme, while annual reports are at the other.

INPUTS

Inputs, including personnel and equipment requirements, can be determined at this point. The resources required to implement the system can be substantial.

SOFTWARE AND HARDWARE

At this stage it is possible to determine the software requirements. Some of the software may be available from commercial sources, while other software requirements will be met via in-house programming. Smaller companies often "farm out" software development to contract programmers.

Finally, after all other items have been determined, the hardware required by the project can be ascertained. One should be careful not to do things backwards, i.e., don't design your system around existing hardware. This is the tail wagging the dog.

Figure 5.3 illustrates the elements of a basic information system. The computer can receive information in a variety of ways. The methods can be divided into two categories: on-line systems and batch systems. On-line systems enter data directly into the computer data base (in real time), while batch systems go through an intermediate stage, such as keypunch, and the data base is updated periodically. Which system to use depends on the use to be made of the data. An inventory tracking system should be an on-line system since people will be using it to locate parts in real time. A quality cost reporting system could be a batch system since it must be up to date only when the monthly report is run. In general, batch systems are less expensive.

With on-line systems, the data are entered directly from the data collection point into the data base. These systems may involve typing in the information at an on-line terminal, reading it in via a bar code sys-

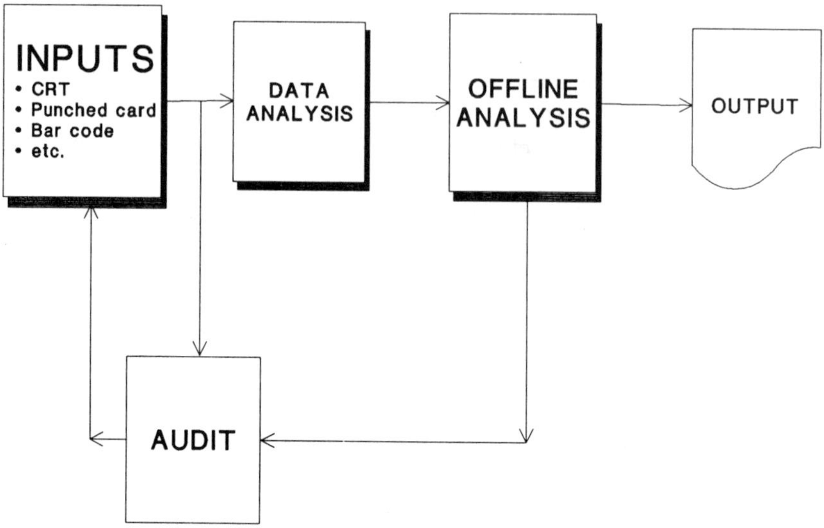

Figure 5.3 Elements of an information system

tem, voice data entry, or input directly from automatic sensors without involving a human being at all. The technology is constantly changing, and it is beyond the scope of this book to cover the subject completely. In selecting the appropriate technology, the system analyst must consider the environment in which the equipment will be used, costs, the amount of data to be collected, and the level of training and number of trained personnel required. Weighing the costs and benefits of the various technologies can be difficult.

Batch systems are generally more paper oriented. Manual systems being converted to computers often go through a batch processing phase on the way to becoming a real-time system. Because most batch systems work from a physical document, the training is directed at first getting the correct information on the document and then getting the information from the document into the computer. The existence of a physical document provides a built-in audit trail and greatly simplifies the task of verifying the accuracy of the information. In the past, transfer of the information from paper into the computer could be done only by having a keypunch operator create a punched card by typing the information all over again. However, new technology includes "smart" optical character readers (OCRs) that can convert information from typed, printed, or even handwritten documents. Computers can also tie in to other computers over telephone lines or through direct links, and radio transmission of computer signals is also possible. As with real-time data entry, the pace of change requires that you research the technology at the time of your project.

More important than the technology is the *accuracy* of the input. The computer trade coined the phrase "garbage in garbage out," or GIGO, to describe the common problem of electronically analyzing inaccurate input data. Accuracy can be improved in a variety of ways. Choosing the appropriate input technology is one way Another is to have computer software check inputs for validity, rejecting obvious errors as soon as they are entered. However,no matter how much input checking takes place, it is usually necessary to have someone audit the input data to ensure accuracy.

DATA PROCESSING

When you are confident that the input data are accurate, you are ready to have the computer analyze the data. Whether to do the analysis on a microcomputer, a minicomputer, a mainframe computer, or a supercomputer is a question of balancing the need for speed and accuracy against the cost. You must also consider the need for other information stored elsewhere in the system, such as the need to access the part number data base. Again, the pace of change makes it essential that you review this question at the time you are designing your system. The mainframe of yesterday is on everyone's desk today.

OFF-LINE ANALYSIS

A truism of quality is that you can never get all of the defects out of the system. A parallel truism of data processing is that "all software has bugs." The result of these two things in combination is the production of computer reports that contain ridiculous errors. Thus, before distribution of any computer report, you should look it over for glaring mistakes. Is the scrap total larger than the sales of the company for the year? Are people who left the company last year being listed as responsible for current activities? Are known problem parts not listed in the top problems? This type of review can save you much embarrassment and can also protect the integrity of the information system you worked so hard to create.

REPORTING

Reports should be sent only to those who need them, and people should be sent only what they need. These simple guidelines are often ignored, creating an information overload for management. As a rule, the level of detail required in a report is greatest at the lower levels of the organization and least at the top. A storeroom clerk may need a complete printout of all the items in the storeroom, breaking out the number of parts in each location. The company president needs to know only the total investment in inventory and the trend. It is recommended that charts be used to condense and summarize large quantities of tabular data when possible.

The distribution of reports should be carefully monitored. From time to time, report recipients should be asked if they still need the report. Also, the report recipients are your *customers*, and their input regarding changes to the reports should be actively solicited.

All of the principles above come together when a QIS is designed and implemented. Few people have the opportunity to create a QIS from scratch. I was fortunate to have had the opportunity to do so. Although I was but one of many players, the experience was quite valuable.

The company involved was growing at a fantastic rate. Within a period of 5 years their employment skyrocketed from 350 people to 1600. As the company grew, their old systems became obsolete. Because of the rapid growth, most of their systems collapsed at the same time. This situation was, of course, a crisis for the company. However, it was also an opportunity to build new systems from the ground up. The case study documents the process of designing these systems from the perspective of a member of quality management.

CASE STUDY *

Not long ago our quality assurance (QA) program used *no* computer time, absolutely none. Yet our manufacturing business was skyrocketing toward $100 million and our product was becoming increasingly complex. Today we utilize computers intensively, including large in-house computers (IBM 370/135/145), time-sharing, and a microcomputer system. We have the ability to determine reject rates for any part, machine, workstation, department, or supplier. We can determine in-

---

stantly whether a part is on hold, the quantity on hold, what is wrong, what is to be done,who is to do it, and exactly where the parts are in any of our 20-odd storerooms with thousands of individual locations. We know scrap rates on a part-by-part or department-by-department basis.

We have a detailed history on field problems, and in some special cases, we have a complete history by unit on high-problem parts (e.g., life history from assembly line to scrap pile). We can determine instantly which parts have caused the greatest field problems in the past and which are the current "top problems" as measured by a variety of means (dollars, most often claimed, greatest quantity claimed). We can call up on-line current warranty problems for any part and get detailed descriptions of problems encountered.

Each month, management receives the latest tally of warranty and field repair charges broken out by model, summarized, plotted, and measured in terms of dollars per unit, claims per unit, and percent of net sales billed.

Complex statistics are routinely computed on a time-sharing computer or a microcomputer. Sample sizes are kept to a minimum by using the most efficient (as opposed to the simplest) statistical methods. When the computer is the mathematician, one can scarcely notice the difference between simple and complex solutions anyhow, especially on a microcomputer when "coretime" is free. The computer's ability to do flawless manipulations of numbers is also useful in such error-prone operations as sorting. The ability to plot and create histograms makes "picturing" data distributions a snap.

We did not just arrive at where we are; we had to get here. Getting here involved considerable trial and error, interfacing with large numbers of people in many areas of the company, familiarizing ourselves with the new world of the computer. The process (which is far from complete) involved a lot of adjusting and some pain. People must be trained, programs must be debugged, procedures must be developed, implemented, and revised.

We found that the successful computer system is normally preceded by a successful manual counterpart. A manual system can be designed in a way that makes the transition to the computer less painful. At times, because of resource limitations, only part of a given system can be computerized, which can cause considerable difficulty if not properly planned and handled.

The computer's role in modern quality assurance is, like the computer itself, complex and often indispensiable. Yet the journey to the computer is fraught with danger and should not be begun without due appreciation of the pitfalls likely to be encountered along the way. By describing one company's rather typical trip down the road to computers, we hope to cast some light on a number of these pitfalls and make the trip easier for future travelers—and all signs indicate that the number traveling this road is likely to increase greatly in the near future.

I wish I could tell you that in the "early days" of my introduction to computers I could have foreseen that today, some 5 years later, we would be where we are now. But it "just ain't so." Back then our QC program was embryonic, having just grown from three people to six (I was number six). Our biggest problem was learning to communicate with one another from department to department and shift to shift. It has been my experience that most action is motivated by the need (actual or imagined) to solve a problem. For 2 years, as our company and department exploded with growth, our problems seemed distant from the computer. Questions such as "How should we organize?" and "Who should I hire?" and "What authority should be delegated?" did not lend themselves to electronic analysis.

Slowly, things began to change. Even though the rapid growth continued, there was a steady flow of data from the 20 people already employed in quality assurance and the need was felt to glean maximum benefit from these data. Thus was born our first "scrap report," a *manual* compilation of all parts scrapped in a given month. A first, the dollars were calculated by searching a computer listing for the cost per piece and then multiplying this by the number of pieces scrapped. This duty was first performed by QA technicians with the help of a part-time clerk, but it was soon obvious that the technicians were spending far too much time in this clerical task. Added to this was the complaint that the cost figures were inaccurate because the parts were scrapped at full cost, even though many were actually scrapped several operations before completion.

With these concerns in mind, the advertisment for a "canned" computer program that generated scrap reports fairly jumped from the page! There was our salvation! Just as a formality, we sent a copy of our letter requesting information to our data processing department (previous approaches to them had gotten us forms to fill out, steeped in the forbidding jargon of the computer world). Much to our surprise, an in-house systems analyst appeared at the QA office a short time later holding a copy of our letter. Why, he asked, did we want to take such a simple problem to an "outsider"? With extremely little resistance, we agreed to "allow" the scrap report to be run by our data processing department.

It seemed to me then (and still does) that people take more notice of a computerized report than of a manual one. So it was with our spanking new scrap report, and when the managers received the report, they promptly set about devising plans to reduce the costs reported. The early result of this attention was to point out a number of errors and shortcomings in the report itself. For a time, these had to be worked out by manually correcting the computer printout.

This was also the first year we began our program of quality engineering in earnest, and my new duties included the application of statistical analysis to day-to-day problems. Most of the analysis was straightforward and could be handled by calculator. Often I resorted to prepared forms, scatter plots, histograms, and the like. Still, inevitably a problem would come up where only a more advanced technique was suitable. Usually, I would first try to deal with the problem using simpler methods, thus wasting time. However, I would finally be forced to yield to the overwhelming logic of the situation and apply the more advanced techniques manually.

If you have never tried it, you should try sometime to work out a multiple regression fitting two or more independent variables by the least-squares method using only a pocket calculator. You will find several sheets of paper, a well-worn pencil, and, if you are like me, errors in your math! The proper model-building procedures (addition of variables, examination of residuals, transformations, etc.) are very likely to be completely ignored.

The in-house computer was very tight on time, partly because of "core-hogging" engineering problems. My initial request for help with routine statistical analysis met with serious questions in data processing. A programmer found a flowchart for a Monte Carlo simulation reliability analysis program in a handbook; it ran on for several pages. "Would I ever need anything like this?" Well, we were getting involved with a new guidance control. . . .

This led to the suggestion that all engineering programs be cleared from the in-house machine to time-sharing service. After some study, all agreed. The transition was made simpler by assistance from the computer time-sharing service company. In fact, they put the largest engineering programs on their machine before we even subscribed. As for statistics, the computer listing of statistics-related programs in the time-sharing library ran for well over 200 pages!

I was also exposed to training in computer programming with help from the time-sharing company. And, just as important, I became more familiar with the computer world's jargon along with the capabilities of computers. My ambitions grew as I learned not only of time-sharing possibilities but also of a more comprehensive assembly of programs. I researched the quality literature, magazines, case histories from other companies, ASQC literature on quality costs, and anything else I could find regarding computers in quality control. Gradually, the outline of a plan formed in my mind, a plan involving documenting the life history of parts from pilot run to obsolescence.

Before QC was really "ready," an opportunity arose to put part of my plan into effect. At about that time, phase I of an elaborate inventory control plan was entering the systems development stage.

The need for *some* advanced inventory control system was felt to be urgent. As our company grew from some $30 million in sales to nearly $60 million in a very short time, the obsolescence of the old system became glaringly obvious. One of the major duties of first-line supervision was to locate the correct parts for a job. This was increasingly difficult as volume increased and more and more parts were manufactured in-house. Added burdens were created by new product introductions until finally, with several thousand part numbers in existence, the old system collapsed completely. The first-generation replacement system was, like most forerunners of successful computer systems, an ingenious manual system. The basis of the system was the move tag, a simple document that was required before any material was moved or stored. Physical control of the parts within each storeroom was based on carefully partitioning the storeroom into definitive locations identified on each move tag.

Despite the vast improvement made by the new manual system, it was soon obvious that it was, at best, a stopgap measure with a number of limitations. One shortcoming was that 16 storerooms existed, not counting a myriad of work-in-process (WIP) locations, and a part might have locations in several storerooms. Another was the great difficulty involved in tying the manual storeroom system to computerized forecasting and ordering. In addition, there was the advantage to be gained from a computerized system in terms of minimizing inventory levels.

The plan for a *total* inventory control system was too much for our in-house mainframe computer system even without the big engineering programs. A decision had to be made as to which elements should be programmed first. For all the reasons given above, storeroom inventory control was selected. It is interesting that, at first, no provision was made to include QC in the planning. The main reason for the exclusion of QC was the feeling that most reject items were from WIP, an element due to be computerized in that never-never land known as "down the road. . . ." When confronted with the real facts (manually obtained), it was obvious that, in reality, big *dollars* were on "hold" status in stores—though goods in process represented high volume and many part numbers. I theorize that the mistaken belief resulted from the fact that our production control group was the prime mover, and rejects in process tend to be more dramatic disrupters of production schedules than rejects in stores. Our logic was accepted immediately and our participation in *all* future planning was assured and welcomed.

We approached the planning for on-line reject goods inventory control not only as an element in a larger system of defective material control but also as a part of a much more comprehensive program of *total* inventory control. The temptation to break off from the rest and push for our own freestanding system was great. In fact, we did explore such an option in detail but rejected it for several reasons:

1. True control involved not just a "paper" system but also physical control of reject goods. If, as an autonomous system implies,this meant removal of goods from the routine flow, it also meant that QA would be instantly in the storeroom business. This was felt to be undesirable.
2. Reject goods were often only delayed or diverted, not removed completely, from the routine flow. Thus, some goods might be repaired on the spot, and insistence on their physical segregation seemed counterproductive.
3. Possibly most important, we saw genuine opportunities for effective material control as a part of the larger system.

The enthusiasm exhibited by those participating in the system development was very impressive. Literally dozens of persons were involved over a period of 6 months, with the activities of all being coordinated by two employees assigned full-time coordination responsibility. One was from data processing and the other from production control. Yet, in spite of the sophistication of the spanking new "on-line real-time" computerized storeroom system, the foundation still rested on two factors already in place: committed, enthusiastic people and physical control of the material. The old manual move tag system was intended to supplement the computer-based system; however, there was one exception.

That exception was for rejected goods. The computer system (approximately 50 computer terminals strategically located for the 1600 employees) recorded for all parts just where the parts were, and this location identification included:

1. Which storerooms (of about 20) the parts were in. (Storerooms are identified by a four-digit number, and this was the sort used on the screen when one desired to see the storeroom inventory display by part.)
2. The location of the parts in the storerooms.
3. The identification numbers on the move tags.
4. How many of the parts were
   a. On the tag.
   b. In the location.
   c. In the storeroom.
   d. In stock (total of all storerooms' inventories).
5. When the last transaction was made.

It was item #3 that provided us with the opportunity to get better control of rejected goods. If any goods were rejected, the computer no longer referenced the move tag number. Instead, it referenced the *reject tag number.* As a futher identifier, all such tag numbers were preceded by the letter "R." This grouped them in the sort and separated them from the nonrejected storeroom supply. This made the control of reject items as good as the inventory control system combined with the quality department's control of reject tags. After a short learning period, the accuracy of the inventory control system was demonstrated and, due to an ongoing audit program by production control, we accepted the accuracy as a given. Control of reject tags, however, took a bit longer.

The problem with reject tag control had been growing for some time and had its roots in our paper-based control system. Each reject tag consisted of a cardboard copy attached to the material rejected, an original that was sent first to the "action department" (i.e., the department responsible for reworking or physically scrapping the product) and then to the quality control office, a carbon copy to be used by production and inventory control, and a third carbon for use by cost accounting in preparing the cost-of-quality report. Our problem was that no real incentive existed for the action department to return the original copy to quality control, which would close the feedback loop. Refer to Figure 5.4.

The new inventory control system made our paper system obsolete virtually overnight, giving us the opportunity to improve it. As mentioned above, the new system allowed identification of goods in any storeroom; thus production and inventory control no longer needed their copy of the reject tag. Likewise, cost accounting arranged to have a periodic listing of "negative transactions" (scrap), which made their copy obsolete. The question was what could be done with the two free copies of each reject tag. We (QC) immediately claimed one copy as our record of a reject tag outstanding. In a very short time we were shocked at the size of the outstanding reject tag pile. How much money was tied up in this inventory? I spoke with the vice-president of manufacturing and found that he, too, wondered. After discussions with other managers in manufacturing, it was agreed that this information, properly presented, could provide a useful scheduling tool. "Properly presented" meant a computer report sorted by action department, part number, and last operation completed. This would allow the work to be integrated with work in process. When we found that data processing was temporarily backlogged, we decided to produce the report on our timesharing system. Cards were keypunched in house, but we had to look up part cost information and enter it by hand.

The reaction to the first report of rejects outstanding was nothing short of astounding. Meetings were held at high levels to discuss means of reducing the volume, responsibility for reduction sometimes being rather emotionally assigned to embarrassed managers. We in QC shared the embarrassment, as a large number of outstanding tags were found to be not outstanding at all but filed away in the QC office. We had simply failed to remove the corresponding tag from the list of outstanding tags. About 40% of the rejects outstanding were cleared in the flurry of activity that followed the release of the first report. About half of this, 20%, was "paper" only as we cleared our files and supervisors scoured every desk drawer and file cabinet looking for "lost" reject tags.

With the unrelenting demands to reduce the volume of rejects outstanding, it was only natural that department supervisors wanted updated reports quickly to convince their superiors that results were indeed being achieved. But our report was scheduled to run monthly and

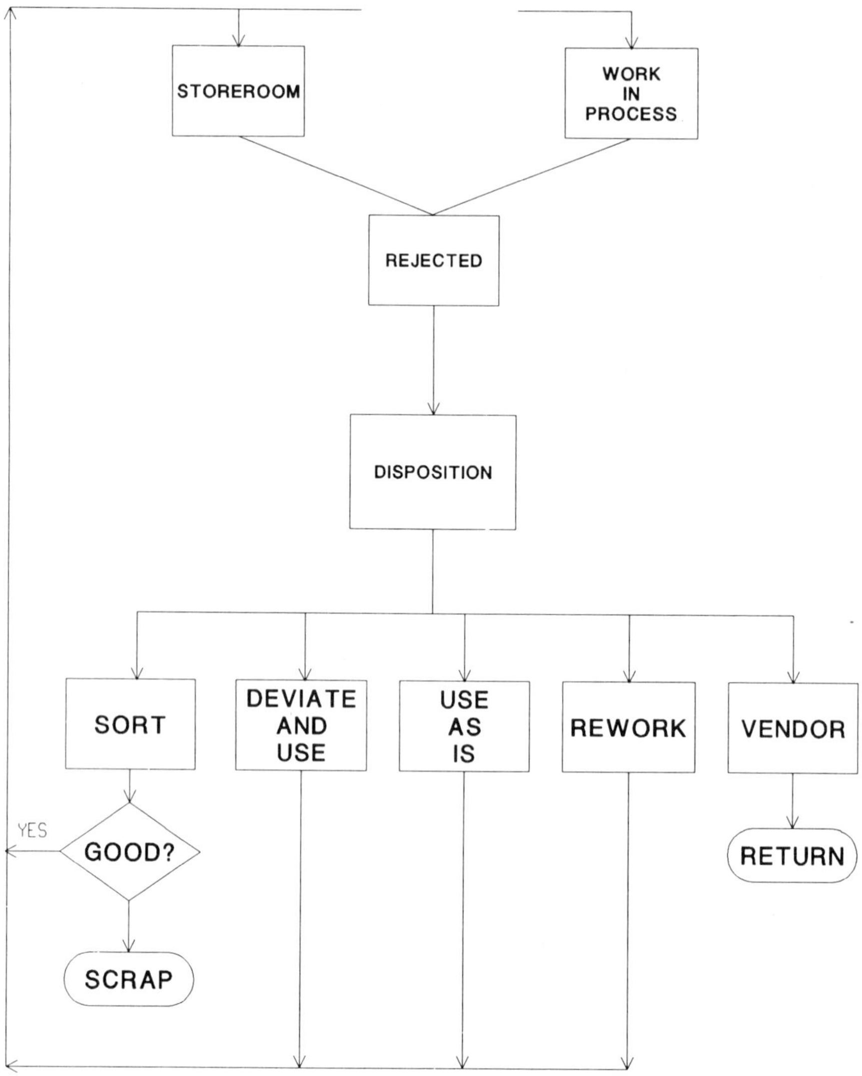

Figure 5.4 Reject material flow.

keypunch scheduling had to be revised. Complaints about lack of timeliness continued to mount until data processing rescheduled and agreed to run the program weekly in-house with data files maintained on line by QC department clerks. We went for several months with good control of rejects outstanding, and we have since returned to monthly reports with the concurrence of manufacturing.

The reports and systems discussed above were still a far cry from the "ultimate system" of internal failure costs, but they were enough of an improvement to move *external failure costs* to the forefront. We began a quiet campaign to convince our warranty group (who reported to the national service manager) that we could help them meet their objectives in reducing warranty costs if they could see their way clear to provide more detailed data on warranty failures. Up to that time the costs had been grouped in very broad categories—nearly the equivalent of "cars and trucks" in the auto industry. But our warranty, like that of most companies, was the accumulation of *part failures*, and we needed to know which parts were the big ones if real progress was to be made. It was finally agreed that a listing of the top ten warranty parts would be included with each monthly report. We would use three different measures: (1) highest dollars year-to-date, (2) highest quantity, and (3) most often claimed. As it turned out, the top ten was both too little and too much: too little because we still lacked data on other parts and still could not fill out our in-house part quality history with accurate field data, and too much because the top ten proved to be formidable problems indeed.

So much was involved in the solution of the top five problems that just the data storage and organization was more than we could handle! Our first response was to generate failure reports on the time-sharing service, but top problems meant frequent demands for the very latest reports and nearly continuous file maintenance. The burden was a real budget buster. Armed with the high cost of the time-sharing reports and the higher cost of field failures, we approached data processing for help. The response exceeded our wildest expectations; after hearing our stories (sometimes long and sad), they offered quality control 36 on-line data files! The files were to be maintained by QC personnel through the on-line network. We chose as our first two files the failure histories of the top two warranty problems. The transition from our manual system of edge-punched cards to the computer system was extremely easy and, shortly, we had fast, accurate data on our top field problems.

With the help of the new data, progress was made toward reducing the magnitude of the problems. Using our third file we put our outstanding reject report on line and made it possible to tie into the on-line storeroom and determine why parts were rejected, which department was responsible, where the parts were located, etc. File 4 was devoted to all parts that required receiving inspection. Any system user can determine whether a part requires inspection, the sampling plans, and the major characteristics to be inspected. File 4 also references any more detailed inspection procedures.

Our quality program also includes periodic part audits, which involve in-depth evaluation of a small sample. We have compiled the results and, over time, this has resulted in a valuable history. The audit history shows not only what parts are rejected but also which ones are accepted, a valuable measure of quality. It also enables us to produce data on the performance of specific machines and processes, measure the performance of the auditors, trace the quality of parts at various operations, and obtain insight into recurring problems.

SUMMARY

The chapter provided a definition of a QIS and its relationship to the broader subject of an MIS. Planning a QIS was described, along with the use of flowcharting as a planning aid. A real-life QIS implementation experience was presented as a case study. The importance of non-computerized elements was explained.

The chapter did not cover any specific hardware or software for QIS. Organizational concerns, such as distributed processing, were not covered. The relationship of QIS to material control systems like materials requirements planning (MRP) or just-in-time (JIT) was not discussed.

RECOMMENDED READING LIST

8–14.

6

Probability and Statistics for Quality Control

The subjects of probability and statistics cover a tremendous amount of material. An individual could easily devote his or her life to either subject without ever exploring it entirely. This chapter will provide the reader with an overview of the elements of the subject critical to understanding basic quality control applications. For further information the reader is referred to one of the many fine books on the subject listed in the Recommended Reading at the end of this book.

One of the most important problems facing anyone concerned with quality improvement is that all processes, no matter how well controlled, exhibit variation. The variation has as its source literally dozens of potential causes. Figure 6.1 shows a cause-and-effect diagram for a particular type of soldering defect, no solder.This particular diagram, based on an actual cause-and-effect diagram used in a statistical process control project, graphically displays only a few of the causes of this solder defect. Still, as you can see, there are dozens.

In a later chapter you will learn to construct cause-and-effect

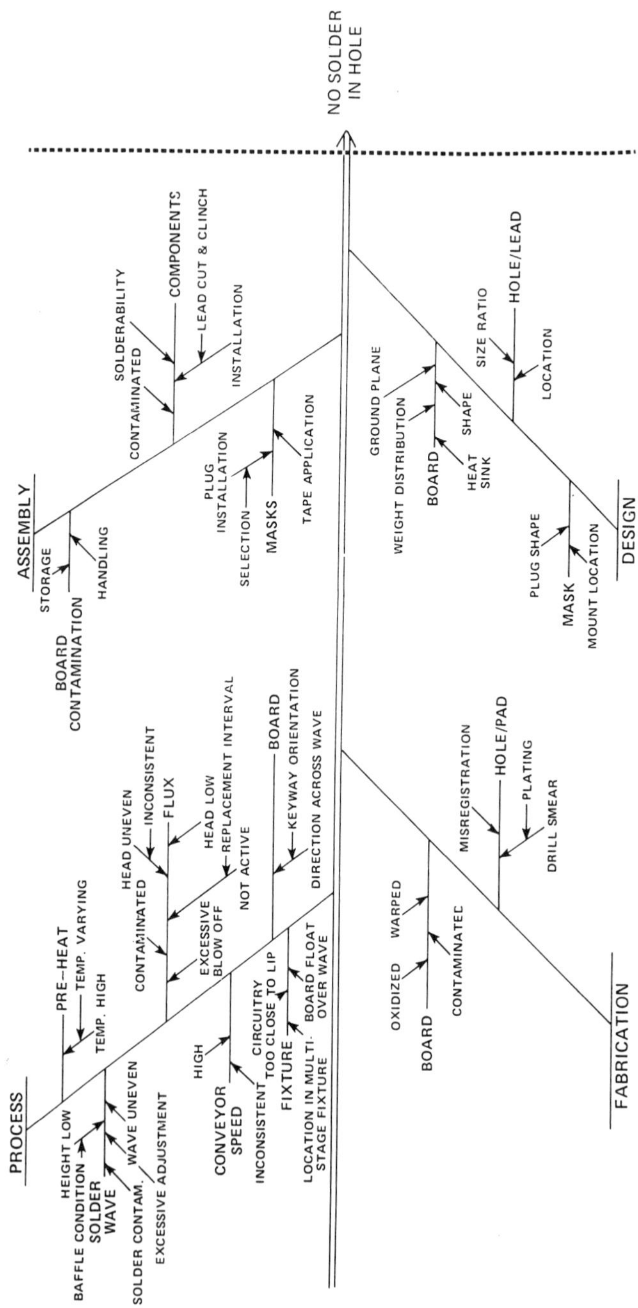

Figure 6.1 Cause of a solder defect.

diagrams. For now, the important point is that any given "effect" is the result of a very large number of causes.

Since we cannot realistically expect zero variation, and since there appear to be too many variables to keep track of, how are we to proceed in our improvement efforts? The answer lies in the use of statistical methods that allow us to understand variability and variation. With this understanding we can determine the appropriate course of action. This chapter describes the basic concepts of probability and statistics that are useful in quality control.

PROBABILITY

Probability theory forms the basis of statistical analysis. Probability is defined as follows:

Definition. Probability is a set function P that assigns to each event A in sample space S a number $P(A)$ such that

1. $P(A)$ is equal to or greater than 0.
2. $P(S) = 1$.
3. If $A_1, A_2, \ldots$ are events that are mutually exclusive [$P(A_i \text{ and } A_j) = 0$], then $P(A_1 \text{ or } A_2 \text{ or} \cdots) = P(A_1) + P(A_2) + \cdots$.

This rather academic definition can be made easier to understand with a simple example. Say I have a penny and I toss it in the air. There are two possible events: it will land with heads up or it will land with tails up. We will be practical here and ignore such extremely unlikely events as the coin landing on an edge or sailing off into space and never landing at all. The first point in the definition is that probability is a set function that assigns a number to each event. The number is called the probability of the event, and it must be at least zero. Futhermore, the sum of all the probabilities must be one. For our case, if the coin is fair, we have the same probability of seeing either a head or a tail. The last condition states that, if the events are mutually exclusive, the probability that one or the other will occur is the sum of the probabilities of the events occurring individually. The events were are considering are mutually exclusive, since we can't have both a head and a tail up at the same time. Thus $P(\text{head}) = .5$ and $\text{P(tail)} = .5$, and $.5 + .5 = 1$.

VENN DIAGRAMS

A simple graphical method that is useful for illustrating many probability relationships and other relationships involving sets is the Venn diagram. Figure 6.2 is Venn diagram of the example we have been discussing.

We say events A_1 and A_2 are *independent* if A_1 does not depend on A_2 and vice versa. When this is the case, the probability of two events occurring simultaneously is simply the product of the probabilities of the events occurring individually. For example, if 5% of our parts are defective, and if we purchase 50% of all our parts from vendor A and 50% of all the *defective* parts are from vendor A, then the probability of a part being both defective and from vendor A is

$$.05 \times .5 = .025$$

Figure 6.2 shows two circles that do not overlap, which means that they are mutually exclusive. If it is possible to get *both* events simultaneously, we have a different situation. Let's say for example that we have a group of 1000 parts (the sample space). Let event A_1 be that a part is defective and event A_2 be that the part is from vendor A. Since it is possible to find a defective part from vendor A, the two events can occur simultaneously. The Venn diagram in Figure 6.3 illustrates this.

If we have the situation illustrated by Figure 6.3, then we can no longer find the probability of either event occurring by just adding the individual probabilities. Instead, we must use the equation

$$P(A_1 \text{ or } A_2) = P(A_1) + P(A_2) - P(A_1 \text{ and } A_2)$$

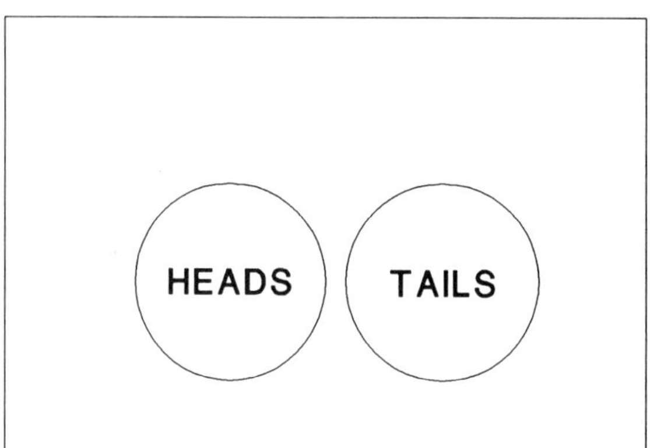

Figure 6.2 Venn diagram of two mutually exclusive events.

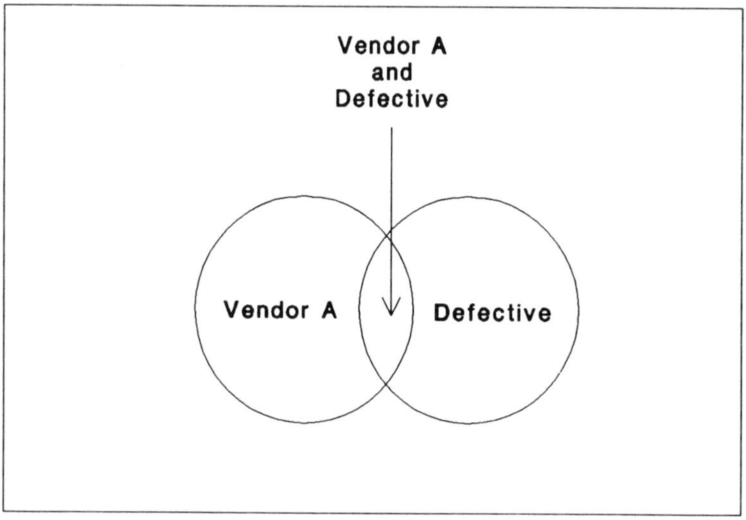

Figure 6.3 Venn diagram of two events that can occur simultaneously.

Returning to our previous example, let A_1 be the event that a part is defective and A_2 be the event that the part is from vendor A. Then, assuming the events are independent, the probability that the part is either defective or from vendor A is

$$.05 + .50 - (.05 \times .50) = .55 - .025 = .525$$

DISCRETE PROBABILITY DISTRIBUTIONS

There are two basic categories of probability distributions, discrete and continuous. A discrete variable is one that can take on only certain values; for example, a count can be only a nonnegative integer. A continuous variable can take on an infinite number of values, at least theoretically. For example, a temperature can be measured to as many decimal places as our measurement instrument can handle. Quality control work involves both discrete and continuous variables.

In applying probability theory to discrete variables in quality control, we frequently encounter the need for efficient methods of counting. One counting technique that is especially useful is combinations. The combination formula is

$$C_r^n = \frac{n!}{r!\,(n-r)!} \tag{1}$$

In the equation above, $n!$ (called "n factorial") means to multiply $n \times (n-1) \times (n-2) \times \cdots (n-n+1)$. Zero factorial is defined as 1 (i.e., $0! \equiv 1$). Equation one gives the number of unique ways in which you can arrange n objects taking them in groups of r objects at a time, where r is a nonnegative integer less than or equal to n. For example, to determine the number of combinations we can make with the letters X, Y, and Z in groups of two letters at a time, we note that $n = 3$ letters, $r = 2$ letters at a time, and use the above formula to find

$$C_2^3 = \frac{3!}{2!\,(3-2)!} = \frac{3 \times 2 \times 1}{(2 \times 1) \times (1)} = \frac{6}{2} = 3$$

The three combinations are XY, XZ, and YZ. Notice that this method does not count reversing the letters as separate combinations; that is, XY and YX are considered to be the same.

SAMPLING WITHOUT REPLACEMENT

Example. Let's look at a practical example of how combinations are applied to quality control. Assume we have received a lot of 12 parts from a distributor. We need the parts badly and are willing to accept the lot if it has fewer than three noncomforming parts. We decide to inspect only four parts since we can't spare the time to check every part. Checking the sample, we find one part that doesn't conform to our requirements. What is the probability that we would find one or fewer if the lot actually had three nonconforming parts?

The formula we need is known as the hypergeometric probability distribution. It is

$$P(x) = \frac{C_{n-x}^{N-m} C_x^m}{C_n^N} \qquad \begin{array}{c} x \leqslant m \\ 0 \leqslant n \leqslant N \\ 0 \leqslant x \leqslant n \end{array} \tag{2}$$

In this equation N is the lot size, m is the number of defectives in the lot, n is the sample size, x is the number of defectives in the sample, and $P(x)$ is the probability of getting exactly x defectives in the sample. We must solve the above equation for $x = 0$ as well as $x = 1$, since we would also accept the lot if we had no defectives. The solution is shown below.

$$P(0) = \frac{C_{4-0}^{12-3}C_0^3}{C_4^{12}} = \frac{126 \times 1}{495} = .255$$

$$P(1) = \frac{C_{4-1}^{12-3}C_1^3}{C_4^{12}} = \frac{84 \times 3}{495} = \frac{252}{495} = .509$$

Adding the two probabilities tells us that probability the our "sampling plan" will accept lots of 12 with 3 nonconforming units. The plan of inspecting four parts and accepting the lot if we have 0 or 1 nonconforming has a probabiltiy of .255 + .509 = .764, or 76.4%, of accepting this "bad" quality lot. This is the "consumer's risk" for this sampling plan. Such a high sampling risk would be unacceptable to most people. The evaluation of sampling plans is taken up in more detail in Chapter 8.

SAMPLING WITH REPLACEMENT

The approach discussed above is exact for evaluating sampling risks when each unit sampled is not returned to the lot before selecting the next sample unit. In practice, sample units are almost always selected at one time; therefore the above approach is correct for nearly every sampling situation involving lots or batches. However, it is not the correct method when sampling from a process in continuous production. In this case we are essentially sampling with replacement, since the "lot size" is the potential production, which is conceptually infinite. Assume that the process is producing some proportion of nonconforming units, which we will call p. If we are basing p on a sample, we find p by dividing the number of nonconforming units by the number of units sampled. The equation that will tell us the probability of getting x defectives in a sample of n units is

$$P(x) = C_x^n p^x (1 - p)^{n-x} \qquad 0 \leqslant x \leqslant n \tag{3}$$

Equation (3) is known as the *binomial probability distribution*. In addition to being useful as the exact distribution of nonconforming units for processes in continuous production, it is an excellent approximation to the hypergeometric probability distribution when the sample size is less than 10% of the lot size.

Example. A process is producing glass bottles on a continuous basis. Past history shows that 1% of the bottles have one or more flaws. If we draw a sample of 10 units from the process, what is the probability that there will be 0 nonconforming bottles?

Using the above information, $n = 10, p = .01$, and $x = 0$. Substituting these values into Equation (3) gives us

$$P(0) = C_0^{10}p^0(1 - .01)^{10-0} = 1 \times 1 \times .99^{10}$$
$$= .904 = 90.4\%$$

Another way of interpreting the above result is that the sampling plan "inspect 10 units, accept the process if no nonconformances are found" has a 90.4% probability of accepting a process that is averaging 1% nonconforming units.

OCCURRENCES PER UNIT

Often in quality control we are not just concerned with *units* that don't conform to requirements; instead we are concerned with the number of nonconformances themselves. For example, say we are trying to control the quality of a computer. A complete audit of the finished computer would almost certainly reveal some nonconformances, even though they might be of minor importance (for example, a decal on the back panel might not be perfectly straight). If we tried to use the hypergeometric or binomial probability distribution to evaluate sampling plans for this situation, we would find they didn't work, because our lot or process would be composed of 100% nonconforming units. Obviously, we are interested not in the units per se, but in the nonconformances themselves.

The correct probability distribution for evaluating nonconformances is the *Poisson distribution*. The equation is

$$P(x) = \frac{u^x e^{-u}}{x!} \qquad x \geqslant 0 \tag{4}$$

In this equation u is the average number of nonconformances per unit, x is the number of nonconformances in the sample, and e is the constant 2.7182818; $P(x)$ gives the probability of exactly x occurrences in the sample.

Example. A production line is producing guided missiles. When each missile is completed, an audit is conducted by an Air Force representative and every nonconformance to requirements is noted. Even though any major nonconformance is cause for rejection, the prime contractor wants to control minor nonconformances as well. Such minor problems as blurred stencils, and small burrs are recorded during the audit. Past history shows that on the average each missile has three minor nonconformances. What is the probability that the next missile will have 0 nonconformances?

We have $u = 3$, $x = 0$. Substituting these values into Equation (4) gives

$$P(x = 0) = \frac{u^x e^{-u}}{x!} = \frac{3^0(2.7182818)^{-3}}{0!} = \frac{1 \times .050}{1} = 0.05$$

In other words, 100% − 5% = 95% of the missiles will have at least one nonconformance.

The Poisson distribution, in addition to being the exact distribution for the number of nonconformances, is a good approximation to the binomial distribution in certain cases. To use the Poisson approximation, you simply let $u = np$ in Equation (4). Juran and Gryna (1980) recommend considering the Poisson approximation if the sample size is at least 16, the population size is at least 10 times the sample size, and the probability of occurrence p on each trail is less than .1. The major advantage of this approach is that it allows you to use the tables of the Poisson distribution, such as Table 2 in the Appendix. Also, the approach is useful for designing sampling plans. The subject of sampling plans is taken up in detail in Chapter 8.

CONTINUOUS PROBABILITY DISTRIBUTIONS

By far, the most common continuous distribution encountered in quality control work is the normal distribution. The probability density function for the normal distribution is given by the equation

$$y(x) = \frac{1}{\sigma\sqrt{2\pi}} e^{-(x-\mu)^2/2\sigma^2} \qquad -\infty < x < \infty \qquad (5)$$

In Equation (5), μ is the population average or mean and σ is the population standard deviation. The population mean is given by

$$\mu = \frac{1}{n} \sum_{i=1}^{n} x_i \tag{6}$$

where the symbol Σ means "to sum." The mean is a measure of the central tendency of the data set. Let's say that we have the following values:

i	x_i
1	17
2	23
3	5

Then the mean is simply the sum of the three values divided by 3, the number of values, or

$$\mu = \frac{1}{3}(17 + 23 + 5) = \frac{45}{3} = 15$$

The standard deviation σ is computed as the square root of the variance σ^2 with the equations

$$\sigma^2 = \frac{1}{n} \sum_{i=1}^{n} (x_i - \mu)^2 \tag{7}$$

$$\sigma = \sqrt{\sigma^2} \tag{8}$$

The variance and standard deviation are both measures of dispersion or spread. Referring to the sample data above with a mean μ of 15, we compute σ^2 and σ as follows:

i	x_i	$x_i - \mu$		$(x_i - \mu)^2$
1	17	2		4
2	23	8		64
3	5	−10		100
			Sum	168

$$\sigma^2 = \frac{168}{3} = 56$$

$$\sigma = \sqrt{56} = 7.483$$

Usually we have only a sample and not the entire population. A *population* is the entire set of observations from which the *sample*, a subset, is drawn. In this case μ is computed in the same way as shown above, but the variance and standard deviation are computed using the formulas below.

$$s^2 = \frac{1}{n-1} \sum_{i=1}^{n} (x_i - \bar{X})^2 \tag{9}$$

$$s = \sqrt{s^2} \tag{10}$$

The differences between these two equations and those used for calculating the population variance and standard deviation are:

1. The Greek letter σ has been changed to s and μ has been changed to an X with a bar over it (pronounced X bar). Generally, Greek letters are used to represent population parameters and English letters are used to represent the sample statistics which estimate these parameters. The quantity X bar is computed in exactly the same way as μ.
2. The denominator of the variance equation has been changed from n to $n - 1$; $n - 1$ is known as the degrees of freedom. Using $n - 1$ makes the statistic s an unbiased estimator of the parameter σ. This simply means that the expected value of $s = \sigma$.

A complete discussion of statistical estimators is beyond the scope of this book. The subject is discussed in most introductory texts on mathematical statistics.

If we plotted a curve using Equation (5) for all values of x we would get a curve like the one shown in Figure 6.4. The normal distribution curve is sometimes called the bell curve because of its shape. This curve is said to be *symmetrical*, meaning that if a line is drawn at the mean of the distribution, the shape of the curve is the same on both sides of the line. It is also said to be *unimodal* because it has only one peak. The areas under the normal curve can be found by integrating Equation (5) (no closed-form solution exists to Equation 5), but more commonly tables are used. Table 1 in the appendix contains areas under the nor-

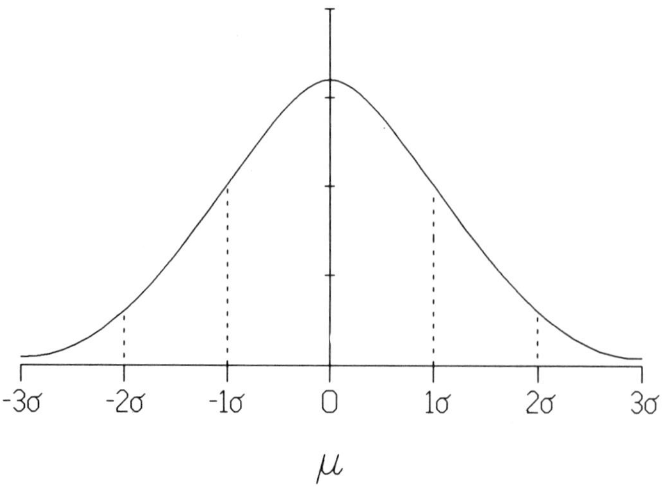

Figure 6.4 Normal distribution.

mal curve. The table is indexed by using the Z transformation, which is

$$Z = \frac{x_i - \mu}{\sigma} \tag{11}$$

By using the Z transformation we convert any normal distribution into a normal distribution with the mean of 0 and a standard deviation of 1. Thus we can use a single normal table to find probabilities.

Example. The normal distribution is very useful in predicting long-term process yields. Assume we have checked the breaking strength of a gold wire bonding process used in microcircuit production and we have found that the process average strength is 9# and the standard deviation is 4#. The process distribution is normal. If the engineering specification is 3# minimum, what percentage of the process will be below the low specification?

We must first compute Z using Equation (11):

$$Z = \frac{3 - 9}{4} = \frac{-6}{5} = -1.2$$

Entering Table 1 for $Z = -1.2$, we find that 11.51% of the area is below this Z value. Thus 11.51% of our breaking strengths will be below our low specification limit of 3#. In quality control applications we usually try to have the average at least 3 standard deviations away from the specification. To accomplish this we would have to improve the process by either raising the average breaking strength or reducing the process standard deviation, or both.

EXPONENTIAL DISTRIBUTION

Another distribution encountered often in quality control work is the exponential distribution. The exponential distribution is especially useful in analyzing reliability. The equation for the probability density function of the exponential distribution is

$$y = \frac{1}{\mu} e^{-x/\mu} \qquad x \geqslant 0 \tag{12}$$

Unlike the normal distribution, the exponential distribution has a highly skewed shape and there is a much greater area below the mean than above it. In fact, over 63% of the expontential distribution falls below the mean. Figure 6.5 shows an exponential distribution curve.

Unlike the normal distribution, the exponential distribution has a closed-form cumulative density function. The equation is shown in the example below.

Example. If a city water company averages 500 system leaks per year, what is the probability that the weekend crew, which works from 6 p.m. Friday to 6 a.m. Monday, will get no calls?

We have $\mu = 500$ leaks per year, which we must convert to leaks per hour. There are 365 days of 24 hours each in a year, or 8760 hours. Thus the mean time between failures (MTBF) is $8760/500 = 17.52$ hours. There are 60 hours between 6 p.m. Friday and 6 a.m. Monday. Thus $x_i = 60$. Using Equation (13), we get

$$P(x \leqslant x_i) = \int_0^{x_i} \frac{1}{\mu} e^{-x/\mu}\, dx = 1 - e^{-x_i/\mu} \tag{13}$$

$$P(x \leqslant 60) = 1 - e^{-60/17.52} = 1 - e^{-3.425} = .967$$

This indicates that 96.7% of the weekends will receive some calls. Thus, the crew will get to loaf away 3.3% of the weekends.

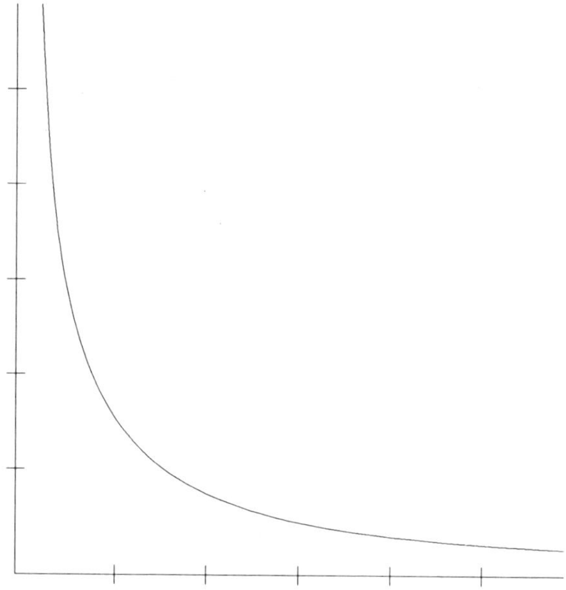

Figure 6.5 Exponential distribution.

STATISTICAL INFERENCE AND HYPOTHESIS TESTING

In almost all practical applications of statistics, including quality control applications, we are forced to make *inferences* about populations based on data from samples. In this chapter we have talked about sample averages and standard deviations; we have even used these numbers to make statements about future performance, such as long-term yields or potential failures. A problem arises that is of considerable practical importance: any estimate that is based on a sample has some amount of sampling error. This is true even if the sample estimates are the "best estimates" in the sense that they are *unbiased* estimators of the population parameters.

Statistical inference generally involves four steps:

1. Formulating a hypothesis about the population or "state of nature"
2. Collecting a sample of observations from the population
3. Calculating statistics based on the sample

4. Either accepting or rejecting the hypothesis based on the predetermined acceptance criterion

There are two types of error associated with statistical inference:

Type I error (α *error*): The probability that a hypothesis that is actually true will be rejected

Type II error (β *error*): The probability that a hypothesis that is actually false will be accepted

Type II errors are often plotted on what is known as an operating characteristics (OC) curve. In subsequent chapters of this book OC curves will be extensively employed in evaluating the properties of various statistical quality control techniques.

So far we have introduced a number of important statistics including the sample mean, the sample standard deviation, and the sample variance. These sample statistics are called *point estimators* because they are but single values used to represent population parameters. It is also possible to construct an interval about the statistics that has a predetermined probability of including the true population parameter. This interval is called a *confidence interval*. Interval estimation is an alternative to point estimation that gives us a better idea of the magnitude of the sampling error.

Confidence intervals are usually constructed as part of a *statistical test of hypotheses*. the hypothesis test is designed to help us make an inference about the true population value at a desired level of confidence. We will look at a few examples of how hypothesis testing can be used in quality control applications. Juran (1979, pp. 22–38) gives a more extensive table and more formal treatment of hypotheses tests.

Example 1. *Experiment*: The nominal specification for filling a bottle with a test chemical is 30 cubic centimeters (cc). The plan is to draw a sample of $n = 25$ units and, using the sample mean and standard deviation, construct a two-sided confidence interval (an interval that extends on either side of the sample average) that has a 95% probability of including the true population mean. If the interval includes 30, conclude that the process mean is 30; otherwise conclude that the process mean is not 30.

Result: A sample of 25 bottles was measured and the following statistics computed:

$$\bar{X} = 28 \text{ cc}$$

$$s = 6 \text{ cc}$$

The appropriate test statistic is t, given by the formula

$$t = \frac{\bar{X} - \mu_0}{s/\sqrt{n}} \tag{14}$$

Substituting our values into the equation we get

$$t = \frac{28 - 30}{6/\sqrt{25}} = \frac{-2}{1.2} = -1.67$$

Table 3 in the appendix gives values for the t statistic at various degrees of freedom. There are $n - 1$ degrees of freedom. For our example we want a two-sided 95% confidence interval, which means that we are willing to accept a 2.5% probability that the true mean will be below our interval and a 2.5% probability that it will be above our interval. This is called a "two-tailed test of hypothesis." Table 3 gives the t values for single-tailed tests, so we need the $t_{.975}$ column, which gives the desired area of 0.025 in the tail, and the row for 24 df. This gives a t value of 2.064. Since the absolute value of this t value is greater than our test statistic, we fail to reject, and therefore accept, the hypothesis that the true mean is 30 cc. In statistical jargon the above example reads like this:

H_0: $\mu = 30$ cc
H_1: μ is not equal to 30 cc
$\alpha = .05$
Critical region: $-2.064 \leqslant t_0 \leqslant +2.064$
Test Statistic: $t = -1.67$

Since t lies in the critical region, fail to reject, and therefore accept, H_0.

Example 2. The variance of machine X's output, based on a sample of $n = 25$, is 100. Machine Y's variance, based on a sample of 10, is 50. The manufacturing representative from the supplier of machine X contends that the result is a mere "statistical fluke." Assuming that a "statistical fluke" is something that has less than 1 chance in 100, test the hypothesis that both variances are actually equal.

The test statistic used to test for equality of two sample variances is the F statistic, given by

$$F = \frac{s_1^2}{s_2^2} \qquad \begin{array}{l} df_1 = n_1 - 1 \\ df_2 = n_2 - 1 \end{array} \tag{15}$$

Substituting our sample values into this equation we get

$$F = \frac{100}{50} = 2 \qquad \begin{array}{l} \text{at 24 df in the numerator} \\ \text{at 9 df in the denominator} \end{array}$$

Using Table 4 for $F_{.99}$ we find that for 24 df in the numerator and 9 df in the denominator $F = 3.61$. Based on this we conclude that the manufacturer of machine x could be right; the result *could* be a statistical fluke. This example demonstrates the volatile nature of the sampling error of sample variances and standard deviations.

Example 3. A machine is supposed to produce parts in the range of 0.500 inch plus or minus 0.006 inch. Based on this, your statistician computes that the absolute worst standard deviation tolerable is 0.002 inch. In looking over your capability charts you find that the best machine in the shop has a standard deviation of 0.0022, based on a sample of 25 units. In discussing the situation with the statistician and management, it is agreed that the machine will be used if a one-sided 95% confidence interval on sigma includes 0.002.

The correct statistic for comparing a sample standard diviation with at standard value is the chi-square statistic. The equation is

$$\text{chi-square} = \frac{(n-1)s^2}{\sigma_0^2} \tag{16}$$

For our data we have $s = 0.0022$, $n = 25$, and $\sigma_0 = 0.002$. The chi-square statistic has $n - 1 = 24$ degrees of freedom. Substituting these values into Equation (16) gives

$$\text{chi-square} = \frac{24(0.0022)^2}{(0.002)^2} = 29.04$$

Table 5 gives, in the .95 column (since we are constructing a one-sided confidence interval) and the df = 24 row, chi-square = 36.42. Since our computed value of chi-square is less than 36.42, we accept the hypothesis and use the machine.

NOTES

This brief introduction into the complex subject of statistics runs a very real risk of leaving you with enough knowledge to be dangerous. If you plan to apply statistical techniques to real-world situations, here are some words of advice:

1. Read the remainder of this book.
2. Take classes and seminars in statistics.
3. Use your common sense in interpreting results.

SUMMARY

This chapter defined probability and described some important probability concepts through the use of Venn diagrams. Sampling with and without replacement was evaluated and quality control examples provided using the hypergeometric and binomial probability distributions. The Poisson distribution and its use in quality control were described. Quality control applications of the normal distribution and exponential distribution were shown. Statistics that measure central tendency and dispersion were shown. A brief introduction to statistical inference and hypothesis testing in a quality control context was provided.

The chapter did not discuss the vast field of mathematical probability and statistics. No derivations or proofs were given for any of the formulas or theorems presented. While many methods of testing distributional assumptions exist, they were not presented. Finally, a wide variety of statistical techniques other than those presented in this chapter exist that can be and have been applied to quality control.

RECOMMENDED READING LIST

41–47, 51–54, 58–59, 61.

7

Statistical Process Control*

Shewhart (1931) defined control as follows:

Definition of Control. A phenomenon will be said to be controlled when, through the use of past experience, we can predict, at least within limits, how the phenomenon may be expected to vary in the future. Here it is understood that prediction within limits means that we can state, at least approximately, the probability that the observed phenomenon will fall within the given limits.

The critical point in this definition is that control is *not* defined as the complete absence of variation. Control is simply a state where all variation is *predictable* variation. A controlled process isn't necessarily a sign of good management, nor is an out-of-control process necessarily producing nonconforming product. In all forms of prediction there is an element of uncertainty. We will call any unknown random cause of

---

*Much of the material in this chapter was taken from *Statistical Process and Quality Control* by Thomas Pyzdek. Copyright 1986. Used by permission of the publisher, Quality America, Inc., Tucson, Arizona.

variation of *chance cause* or a *common cause*; the terms are synonymous and will be used as such. If the influence of any particular chance cause is very small, and if the number of chance causes of variation is very large and relatively constant, we have a situation where the variation is predictable within limits. You can see from our definition above that a system such as this qualifies as a controlled system. Where Dr. Shewhart used the term "chance cause," Dr. W. Edwards Deming coined the term "common cause" to describe the same phenomenon. Both terms are encountered in practice.

Of course, not all phenomena arise from constant systems of common causes. At times the variation is caused by a source of variation that is not part of the constant system. These sources of variation were called *assignable causes* by Shewhart and *special causes* by Deming. Experience indicates that special causes of variation can usually be identified and corrected, leading to a process that is less variable.

Statistical tools are needed to help us identify the effects of special causes of variation. This leads to another definition:

Statistical Process Control. The use of statistical methods to identify the presence of special causes of variation in a process.

The basic rule of statistical process control is:

> *Variation from common cause systems should be left to chance, but special causes of variation should be identified and eliminated.*

Figure 7.1 illustrates the need for statistical methods to determine the category of variation. The figure shows two time-ordered sequences of fractions defective on two different apparatus. Without statistical guidance it is likely that people would disagree on whether the observed variations were due to chance or whether some assignable cause of variation was affecting the process. Human beings have a very poor record of being able to discern the difference between chance variation and caused variation without such guidance.

The answer to the question "Should these variations be left to chance?" can only be obtained through the use of statistical methods. Figure 7.2 illustrates the basic concept.

In short, variation between the two "control limits" designated by the dashed lines in Figure 7.2 will be deemed as variation from the common cause system. Any variability beyond these fixed limits will be assumed to have come from special causes of variation. We will call any system exhibiting only common cause variation a statistically controlled system. It must be noted that the control limits are not simply

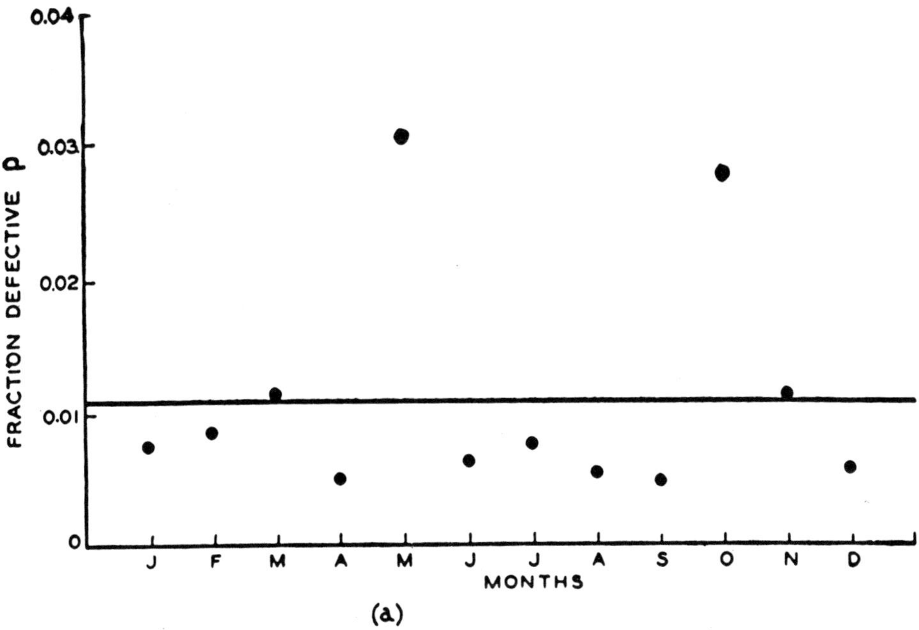

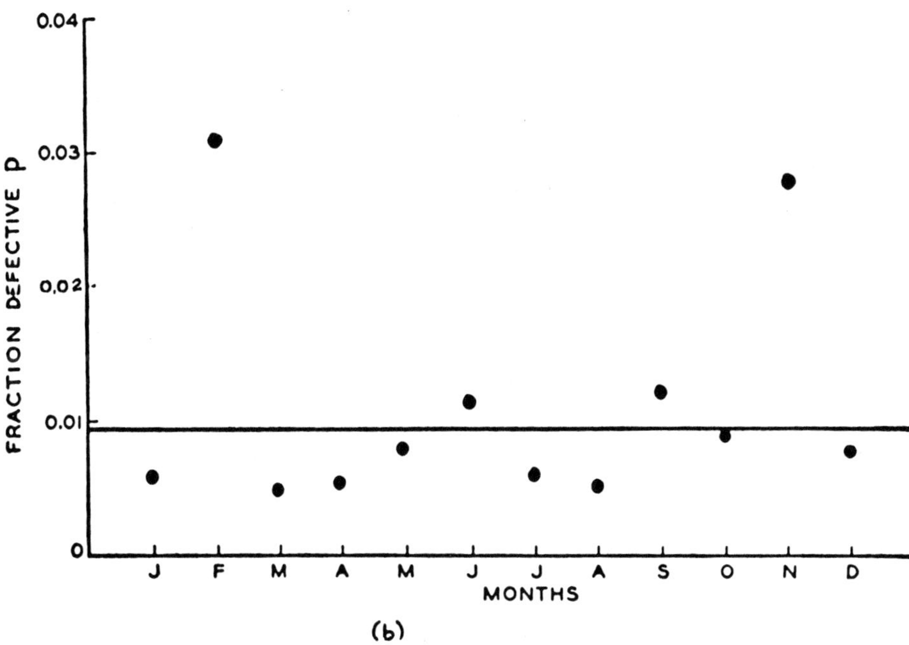

Figure 7.1 Should these variations be left to chance? (From Walter A. Shewhart; *Economic Control of Quality of Manufactured Product*, 1931, p. 13. Used by permission of the publisher, American Society for Quality Control).

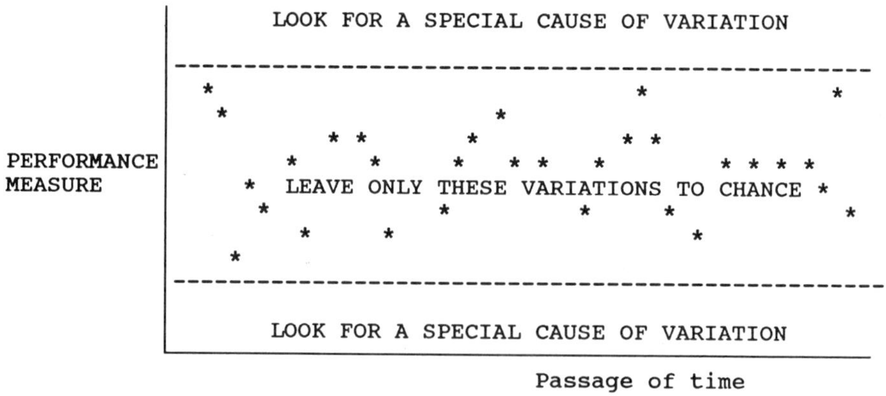

Figure 7.2 Types of Variation.

pulled out of the air; they are calculated from statistical theory. Figure 7.3 shows the control limits drawn on Figure 7.1. Notice that figure 7.3a exhibits variations from special causes, while Figure 7.3b does not. This implies that the types of action needed to reduce the variability are of a different nature. In the case of Figure 7.3a variability reduction should begin with a search for assignable causes of variation, while the situation in Figure 7.3b indicates variability that is arising from chance causes of variation.

DISTRIBUTIONS

A central concept in statistical process control is that almost every measurable phenomenon is a statistical distribution. In other words, an observed set of data constitutes a sample of the effects of unknown common causes. It follows that, after we have done everything to eliminate special causes of variations, there will still remain a certain amount of variability exhibiting the state of control. Figure 7.4 illustrates the relationships between common causes, special causes, and distributions.

There are three basic properties of a distribution: location, spread, and shape. A distribution can be characterized by these three parameters. The location refers to the typical value of the distribution. In the previous chapter we discussed one measure of location, the mean. The spread of the distribution is the amount by which smaller values differ

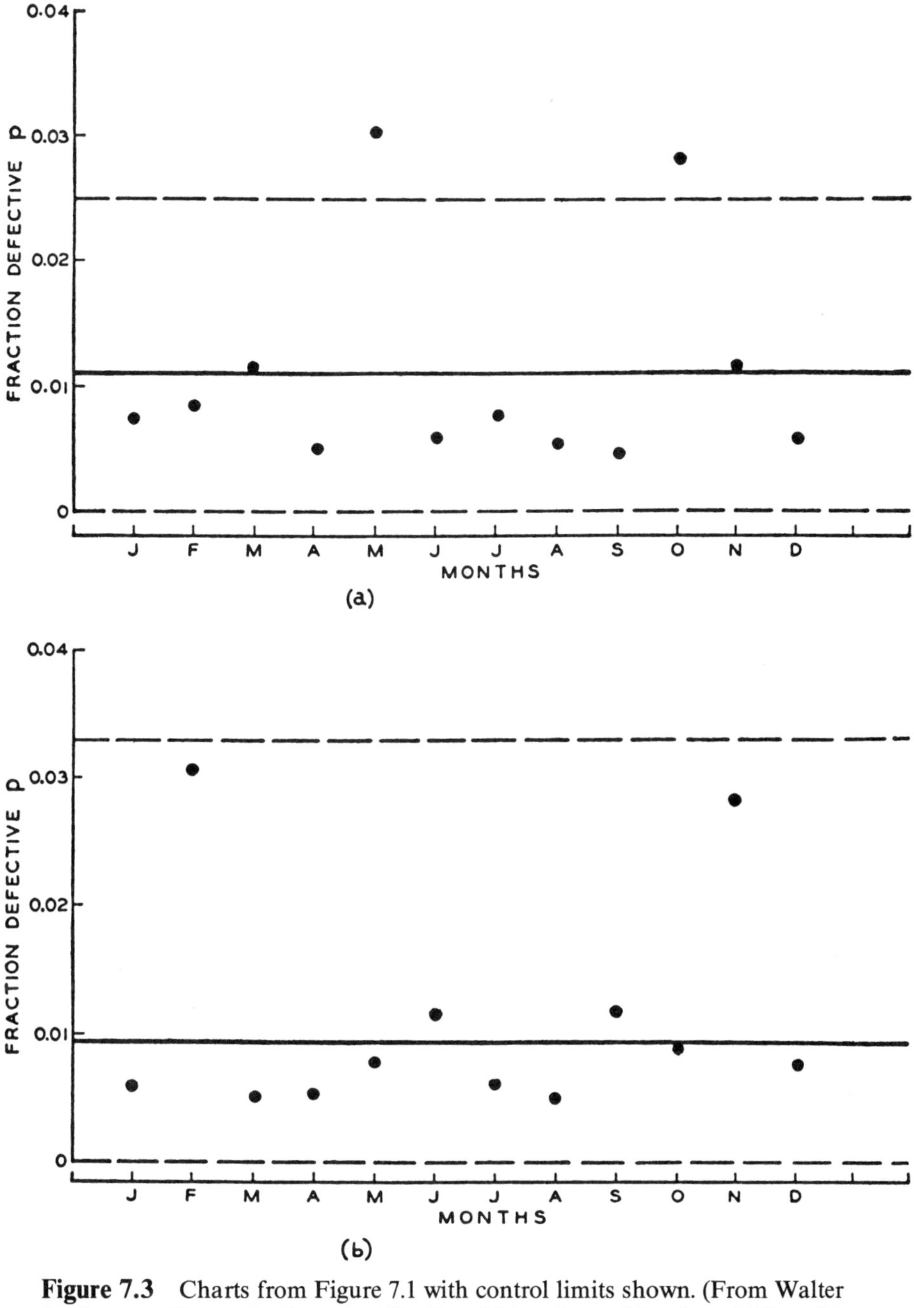

Figure 7.3 Charts from Figure 7.1 with control limits shown. (From Walter A. Shewart, *Economic Control of Quality of Manufactured Product, 1931, p. 16. Used by permission of the publisher, American Society for Quality Control.)*

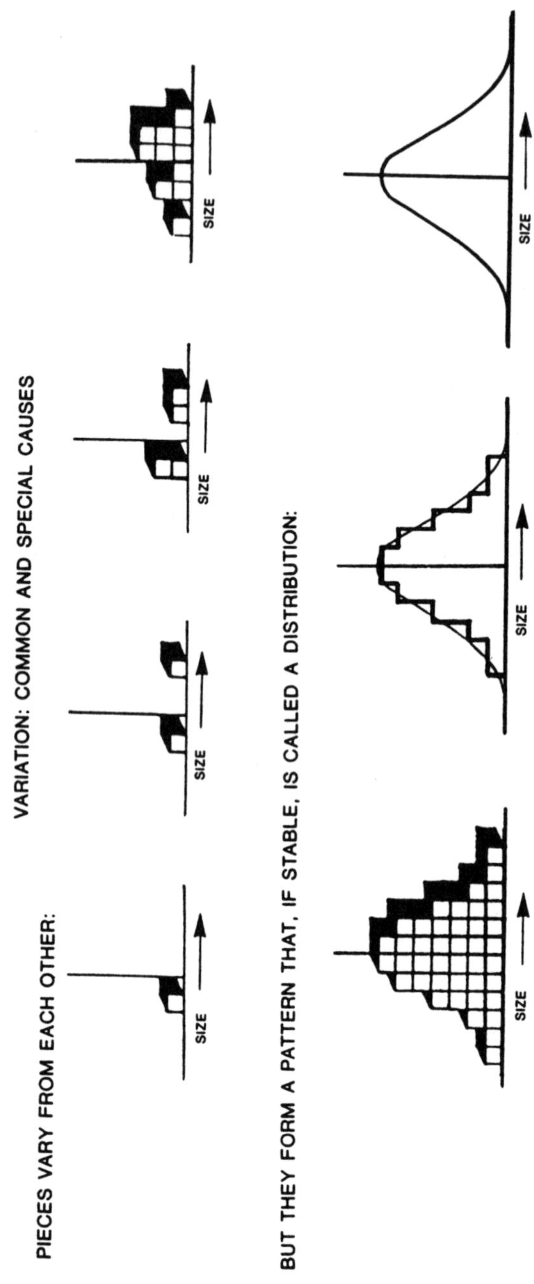
PIECES VARY FROM EACH OTHER:
VARIATION: COMMON AND SPECIAL CAUSES
BUT THEY FORM A PATTERN THAT, IF STABLE, IS CALLED A DISTRIBUTION:
SIZE

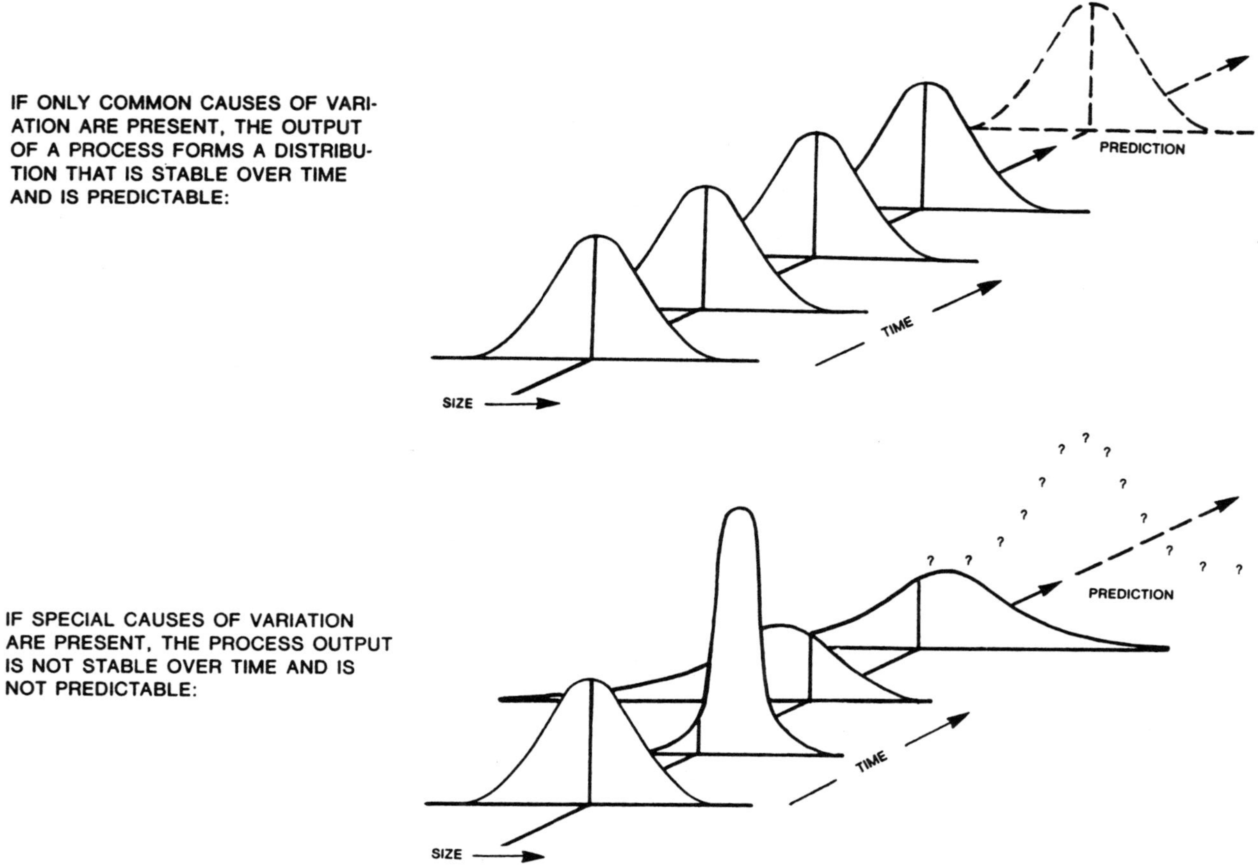

IF ONLY COMMON CAUSES OF VARIATION ARE PRESENT, THE OUTPUT OF A PROCESS FORMS A DISTRIBUTION THAT IS STABLE OVER TIME AND IS PREDICTABLE:
PREDICTION
TIME
SIZE
IF SPECIAL CAUSES OF VARIATION ARE PRESENT, THE PROCESS OUTPUT IS NOT STABLE OVER TIME AND IS NOT PREDICTABLE:
PREDICTION
TIME
SIZE

Figure 7.4 (From *Continuing Process Control and Process Capability Improvement,* p. 4a. Copyright 1983. Used by permission of the publisher, Ford Motor Company, Dearborn, Michigan.)

from larger ones. The standard deviation and variance are measures of distribution spread. The shape of a distribution is its pattern—peakedness, symmetry, etc. A given phenomenon may have any one of a number of distribution shapes (the distribution may be bell-shaped, rectangular, etc.). In the previous chapter we discussed two common continuous distributions with distinctly different shapes, the normal distribution and the exponential distribution.

CENTRAL LIMIT THEOREM

The central limit theorem can be stated as follows:

> Irrespective of the shape of the distribution of the population or universe, the distribution of average values of samples drawn from that universe will tend toward a normal distribution as the sample size grows without bound.

It can also be shown that the average of sample averages equals the average of the universe and that the standard deviation of the averages equals the standard deviation of the universe divided by the square root of the sample size. Shewhart performed experiments that showed that small sample sizes (approximately four or five) were required to get approximately normal distributions from even wildly nonnormal universes. Figure 7.5 was created by Shewhart using samples of four units.

At this point you may be asking, so what? The practical implications of the central limit theorem are immense. Without the central limit theorem you would have to develop a separate statistical model for every nonnormal distribution encountered in practice. This would be the only way to determine whether the system was exhibiting chance variation. Because of the central limit theorem you can use *averages* of small samples to evaluate *any* continuous and many discrete process measurements using the normal distribution. The central limit theorem is the basis for the most powerful of statistical process control tools, Shewhart control charts.

REPRESENTING VARIATION IN A PROCESS

Measures of Central Tendency

As stated above, measurements from a process can be represented by a distribution. The distribution, in turn, can be characterized by its loca-

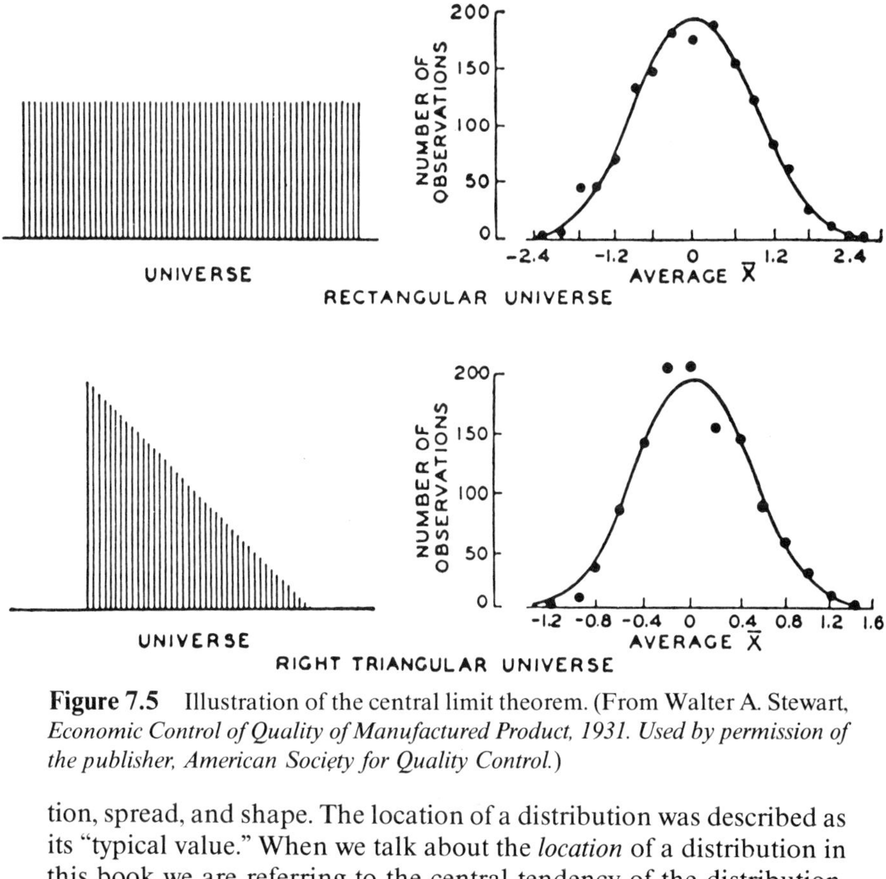

Figure 7.5 Illustration of the central limit theorem. (From Walter A. Stewart, *Economic Control of Quality of Manufactured Product, 1931. Used by permission of the publisher, American Society for Quality Control.*)

tion, spread, and shape. The location of a distribution was described as its "typical value." When we talk about the *location* of a distribution in this book we are referring to the central tendency of the distribution. Remember that we said a distribution is the pattern that remains after we have eliminated all special causes of variation. This assumes that we have obtained a reasonably large number of samples, analyzed their pattern, and found that they exhibit variation that could arise from a common cause system. What might a typical value from this common cause system be? Intuitively, the value that divides the data into two equal parts is a typical value. This is known as the median. The median is calculated as follows.

Steps for finding the median

1. Sort the data from smallest to largest, or largest to smallest.
2. If the count of values is an odd number, then the median is the mid-

dle value. For example, if you have five numbers the median is the third value. The example below illustrates this.

Data: 4.3, 3.2, 1.1, 4.0, 2.4

Rank	Value	
1	1.1	
2	2.4	
3	3.2	← 3rd value = median = 3.2
4	4.0	
5	4.1	

3. If the count of values is an even number, then the median is obtained by dividing the two middle values by 2. For example,

Data: 4.3, 1.1, 4.0, 2.4

Rank	Value	
1	1.1	
2	2.4	these are the middle values
3	4.0	
4	4.1	

$$\text{Median} = \tilde{X} = \frac{2.4 + 4.0}{2} = \frac{6.4}{2} = 3.2$$

Note the symbol for the median, "x tilde."

Another typical value might be the value that occurs most often. The statistical term for this value is the *mode*. Figure 7.6 illustates the mode.

The final typical value we will consider here is the *mean*, or average value. The mean is the center of gravity of a distribution and it is found by adding up all the values and dividing the sum by the count. Mathematically, here is how we write this:

$$\text{Mean} = \bar{X} = \frac{1}{n} \sum_{i=1}^{n} x_i = \frac{\text{sum of values}}{\text{number of values}} \tag{18}$$

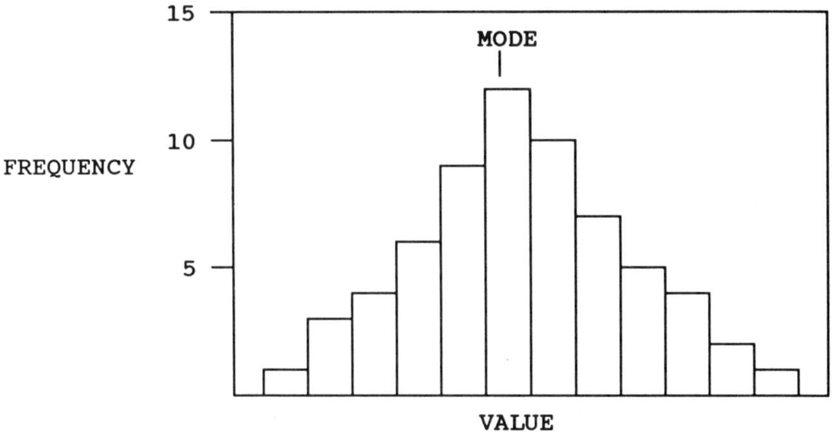

Figure 7.6 The Mode.

The mean is by far the most commonly used typical value, or measure of central tendency of a distribution. You may wish to compare Equation (18), the sample mean, the Equation (6), the population mean.

Measures of Dispersion

The second property of a distribution that we will want to measure is its spread, or dispersion. There are a number of useful statistics that quantify a distribution's spread. The variance and standard deviation, described in the previous chapter, have a number of desireable mathematical properties. However, the easiest measure of dispersion to calculate is the range.

$$\text{Range} = R = \text{largest value} - \text{smallest value} \tag{19}$$

Fortunately, when the distribution is from a constant system of chance causes, that is, the distribution is from a process in a state of statistical control, we can use the average range to obtain a very satisfactory estimate of σ. The relationship is given by

$$\hat{\sigma} = \frac{\bar{R}}{d_2} \tag{20}$$

where d_2 is a constant. Table 6 gives d_2 values for various sample sizes. Please note that Equation (20) holds *if and only if the process is in statistical control.*

PREVENTION VERSUS DETECTION

A process control system is essentially a feedback system. Four main elements are involved: the process itself, information about the process, action taken on the process, and action taken on the output from the process. The way these elements fit together is shown by Figure 7.7.

By the process, we mean the whole combination of people, equipment, input materials, methods, and environment that work together to produce output. The performance information is obtained, in part, from evaluation of the process output. The output of a process includes more than product; it also includes information about the operating state of the process such as temperature and cycle times. Action taken on a *process* is future-oriented in the sense that it will affect output yet to come. Action on the *output* is past-oriented because it involves detecting out-of-specification output that has already been produced.

There has been a tendency in the past to concentrate attention on the past-oriented strategy of inspection. With this approach we wait until output has been produced; then the output is inspected and either

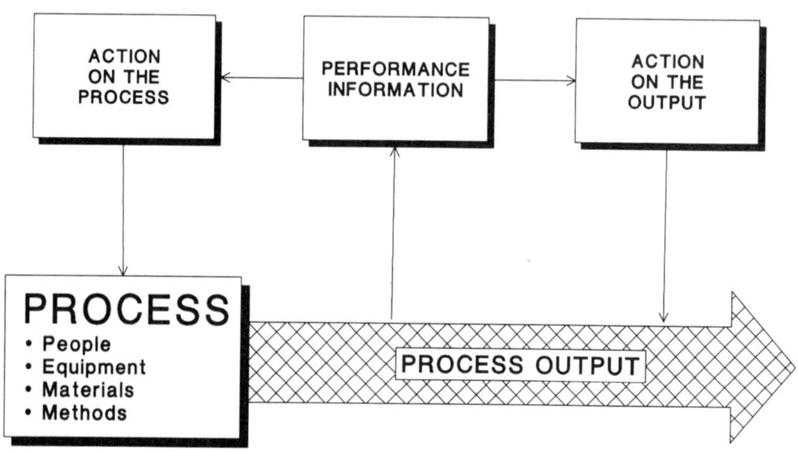

Figure 7.7 A process control system.

accepted or rejected. It should be obvious that this does nothing to prevent substandard output in the future. Statisical process control takes us in a completely different direction: improvement in the future. A key concept is that *the smaller the variation around the target, the better*. Thus, according to this school of thought, it is not enough to merely meet the requirements. Continuous improvement is called for. The concept of never-ending, continuous improvement is at the heart of SPC.

CAUSE-AND-EFFECT DIAGRAMS

Process improvement involves taking action on the causes of variation. With most practical applications the number of possible causes for any problem can be huge. Dr. Kaoru Ishikawa developed a simple method of graphically displaying the causes of any given quality problem. His method is called by several names: the Ishikawa diagram, the fishbone diagram, or the cause-and-effect diagram.

Cause-and-effect diagrams are tools that are used to organize and graphically display all of the knowledge a group has related to a particular problem. Usually, the steps are

1. Develop a flowchart of the area to be improved.
2. Define the problem to be solved.
3. Brainstorm to find all possible causes of the problem.
4. Organize the brainstorming results in rational categories.
5. Construct a cause-and-effect diagram that accurately displays the relationships of all the data in each category.

Once these steps are complete, constructing the cause-and-effect diagram is very simple. The steps are

1. Draw a box on the far right-hand side of a large sheet of paper and draw a horizontal arrow that points to the box. Inside the box, write the description of the problem you are trying to solve.
2. Write the names of the categories above and below the horizontal line. Think of these as branches from the main trunk of the tree.
3. Draw in the detailed cause data for each category. Think of these as limbs and twigs on the branches.

A good cause-and-effect diagram will have many twigs. (See Figure 7.8). If your cause-and-effect diagram doesn't have a lot of smaller branches and twigs, it shows that the understanding of the problem is superficial. Chances are you need the help of someone outside your

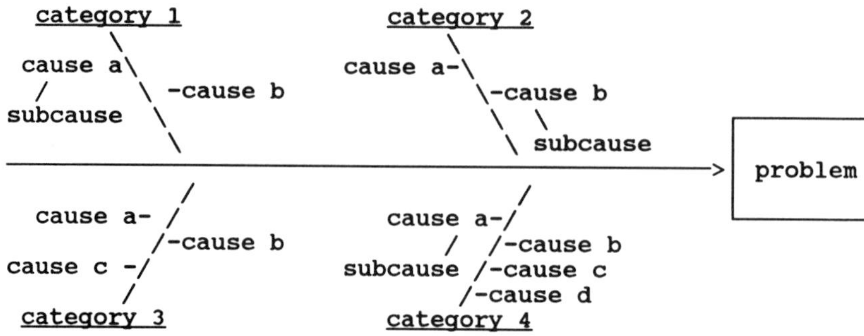

Figure 7.8 Cause-and-effect diagram.

group to aid in the understanding, perhaps someone more closely associated with the problem.

Cause-and-effect diagrams come in several basic types. The *dispersion analysis type* is created by repeatedly asking "Why does this dispersion occur?" For example, we might want to know why all of our fresh peaches don't have the same color.

The *production process class* cause-and-effect diagram uses production processes as the main categories, or branches of the diagram. In Figure 7.9 processes are shown joined by the horizontal line.

The *cause enumeration* cause-and-effect diagram, as shown in Figure 7.8, simply displays all possible causes of a given problem grouped according to rational categories. This type of cause-and-effect diagram lends itself readily to the brainstorming approach.

Cause-and-effect diagrams have a number of uses. Creating the diagram is an education in itself. Organizing the knowledge of the

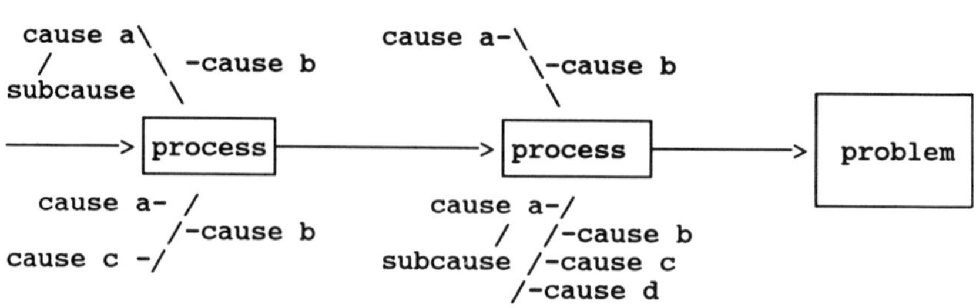

Figure 7.9 Production process class cause-and-effect diagram.

group serves as a guide for discussion and frequently inspires more ideas. The cause-and-effect diagram, once created, acts as a record of your research. Simply record your tests and results as you proceed. If the true cause is found to be something that wasn't on the original diagram, write it in. Finally, the cause-and-effect diagram is a display of your current level of understanding. It shows the existing level of technology as understood by the team. It is a good idea to post the cause-and-effect diagram in a prominent location for all to see.

CEDAC

A variation of the basic cause-and-effect diagram, developed by Dr. Ryuji Fukuda of Japan, is the cause-and-effect diagrams with the addition of cards, or CEDAC. The main difference is that the group gathers ideas outside the meeting room on small cards, as well as in group meetings. The cards also serve as a vehicle for gathering input from people who are not in the group; they can be distributed to anyone involved with the process. Often the cards provide more information than the brief entries on a standard cause-and-effect diagram. The cause-and-effect diagram is built by actually placing the cards on the branches. The reader is referred to Fukuda (1983) for additional information on CEDAC.

DESCRIPTIVE DATA ANALYSIS

Statistical process control, technically, uses only inferential statistical techniques. However, in practice some descriptive techniques have proved very useful in SPC efforts. A descriptive statistical technique merely reorganizes the data without making any inference regarding population parameters. Descriptive techniques are explorative in nature. Two descriptive techniques are described here: Pareto analysis and histograms. The information is presented in a "procedure format" to make it easier to see how the technique is actually used. This format will be used for all the SPC techniques presented in this book.

Pareto Analysis

Definition. Pareto analysis is the process of ranking opportunities to determine which of many potential opportunities should be pursued first. It is also known as "separating the vital few from the trivial many."

Usage. Pareto analysis should be used at various stages in a quality improvement program to determine which step to take next. It is used to answer such questions as "What department should have the next SPC team?" or "On what type of defect should we concentrate our efforts?"

How to Perform a Pareto Analysis

1. Determine the classifications (Pareto categories) for the graph. If the desired information does not exist, obtain it by designing checksheets and logsheets.
2. Select a time interval for analysis. The interval should be long enough to be representative of typical performance.
3. Determine the total occurrences (i.e., cost, defect counts, etc.) for each category. Also determine the grand total. If several categories account for only a small part of the total, group these into a category called "other."
4. Compute the percentage for each category by dividing the category total by the grand total and multiplying by 100.
5. Rank order the categories from the largest total occurrences to the smallest.
6. Compute the "cumulative percentage" by adding the percentage for each category to that of any preceding categories.
7. Construct a chart with the left vertical axis scaled from zero to at least the grand total. Put an appropriate label on the axis. Scale the right vertical axis from 0 to 100%, with 100% on the right side being the same height as the grand total on the left side.
8. Label the horizontal axis with the category names. The leftmost category should be the largest, next the second largest, and so on.
9. Draw in bars representing the amount of each category. The height of the bar is determined by the left vertical axis.
10. Draw on line that shows the cumulative percentage column of the Pareto analysis table. The cumulative percentage line is determined by the right vertical axis.

Example of Pareto Analysis. The following data have been recorded for peaches arriving at Super Duper Market during August.

Problem	Peaches lost
Bruised	100
Undersized	87
Rotten	235
Underripe	9
Wrong variety	7
Wormy	3

The Pareto table and diagram for the above data are presented below.

Rank	Category	Count	Percentage	Cumulative %
1	Rotten	235	53.29	53.29
2	Bruised	100	22.68	75.97
3	Undersized	87	19.73	95.70
4	Other	19	4.31	100.01

Note that, as often happens, the final percentage is slightly different from 100%. This is due to round-off error and is nothing to worry about.

Histograms

Definition. A plot of a frequency distribution in the form of rectangles whose bases are equal to the cell interval and whose areas are proportional to the frequencies. In simpler terms, a histogram is a pictorial representation of a set of data.

Usage. Histograms are used to determine the shape of a data set. Also a histogram displays the numbers in a way that makes it easy to see the dispersion and central tendency and to compare the distribution to requirements. Histograms can be valuable troubleshooting aids. Comparisons between histograms from different machines, operators, vendors, etc. often reveal important differences.

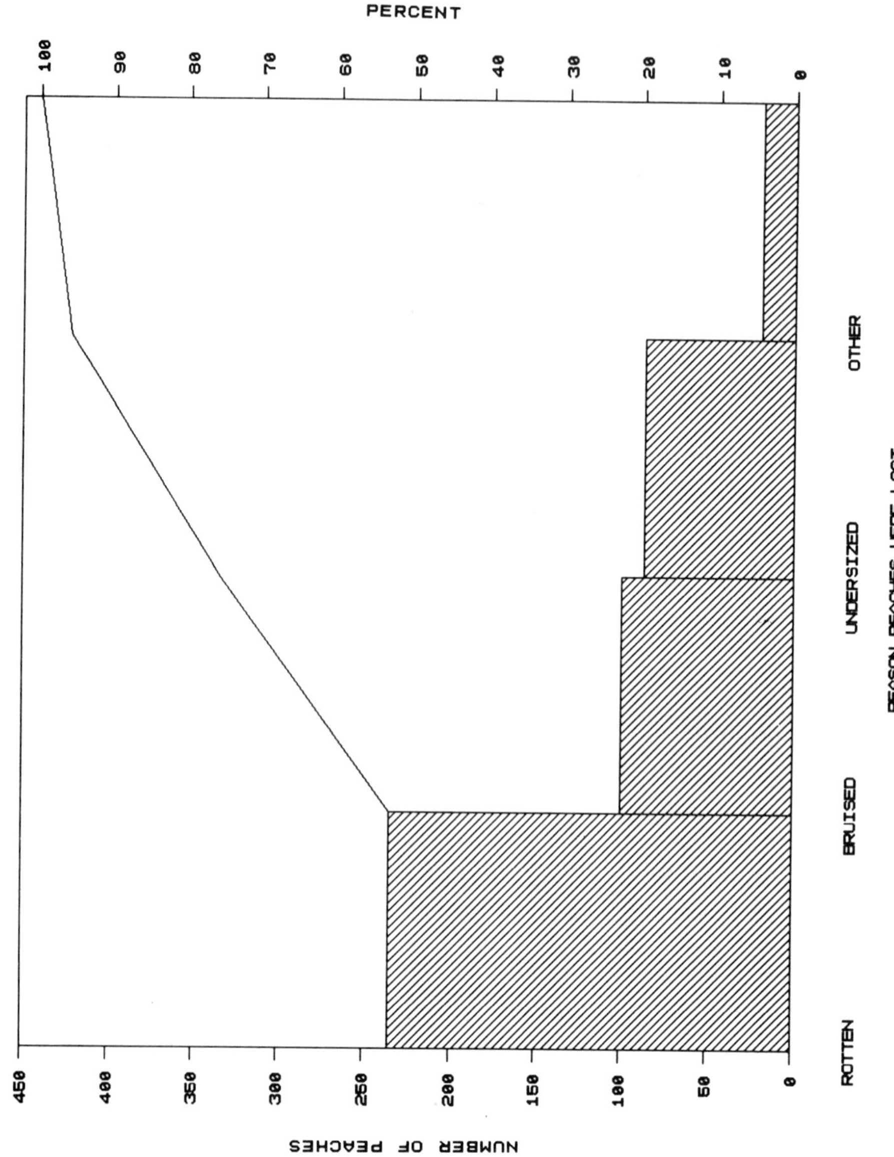

Figure 7.10 Completed Pareto diagram for lost peaches.

How to Construct a Histogram

1. Find the largest and the smallest value in the data.

2. Compute the range (we use the letter R to represent the range throughout this book) by subtracting the smallest value from the largest value. [See Equation (19).]

$$R = \text{largest value} - \text{smallest value}$$

3. Select a number of cells for the histogram. The table below provides some guidelines:

Number of values	Number of cells
100 or less	7–10
101–200	11–15
201 or more	13–20

The final histogram may not have exactly the number of cells you choose here, because of rounding.

4. Determine the width (W) of each cell. The cell width W is computed from the formula

$$W = \frac{R}{\text{number of cells}} \tag{21}$$

5. The number W is a starting point. You should round W to a convenient number.

6. Compute "cell boundaries." A cell is a range of values, and cell boundaries define the start and end of each cell. To avoid ambiguity, cell boundaries should have one more decimal place than the raw data values in the data set. The low boundary of the first cell must be less than the smallest vaue in the data set. Other cell boundaries are found by adding W to the previous boundary. Continue until the upper boundary of a cell is larger than the largest value in the data set.

7. Compute the cell midpoint for the first cell by adding the lower and upper boundaries of the first cell and dividing the total by 2. The midpoint of subsequent cells can be found by adding W to the previous midpoint.

8. Now go through the raw data and determine the cell into which each value falls. Mark a tick in the appropriate cell.

9. Count the ticks in each cell and record the count, also called the frequency, to the right of the tick marks.

10. Construct a graph from the table. The vertical axis of the graph will show the frequency in each cell. The horizontal axis will show the cell boundaries.

11. Draw bars representing the cell frequencies. The bars should all be the same width, and the height of the bars should equal the frequency in the cell.

Histogram Example. Assume you have the following data on the size of a metal rod.

Row	Data				
1	1.002	0.995	1.000	1.002	1.005
2	1.000	0.997	1.007	0.992	0.995
3	0.997	1.013	1.001	0.985	1.002
4	0.990	1.008	1.005	0.994	1.012
5	0.992	1.012	1.005	0.985	1.006
6	1.000	1.002	1.006	1.007	0.993
7	0.984	0.994	0.998	1.006	1.002
8	0.987	0.994	1.002	0.997	1.008
9	0.992	0.988	1.015	0.987	1.006
10	0.994	0.990	0.991	1.002	0.988
11	1.007	1.008	0.990	1.001	0.999
12	0.995	0.989	0.982*	0.995	1.002
13	0.987	1.004	0.992	1.002	0.992
14	0.991	1.001	0.996	0.997	0.984
15	1.004	0.993	1.003	0.992	1.010
16	1.004	1.010	0.984	0.997	1.008
17	0.990	1.021*	0.995	0.987	0.989
18	1.003	0.992	0.992	0.990	1.014
19	1.000	0.985	1.019	1.002	0.986
20	0.996	0.984	1.005	1.016	1.012

The frequency table for the above data is shown below and the histogram is given in Figure 7.11.

Cell	Cell boundaries	Midpoint	Tally	Frequency
1	0.9815–0.9855	0.9835	\\\\\\\\	8
2	0.9855–0.9895	0.9875	\\\\\\\\\	9
3	0.9895–0.9935	0.9915	\\\\\\\\\\\\\\\\\	17
4	0.9935–0.9975	0.9955	\\\\\\\\\\\\\\\\	16
5	0.9975–1.0015	0.9995	\\\\\\\\\	9
6	1.0015–1.0055	1.0035	\\\\\\\\\\\\\\\\\\\	19
7	1.0055–1.0095	1.0075	\\\\\\\\\\\	11
8	1.0095–1.0135	1.0115	\\\\\\	6
9	1.0135–1.0175	1.0155	\\\	3
10	1.0175–1.0215	1.0195	\\	2

Average and Range Charts

Definition. Average charts are statistical tools used to evaluate the central tendency of a process over time. Range charts are statistical

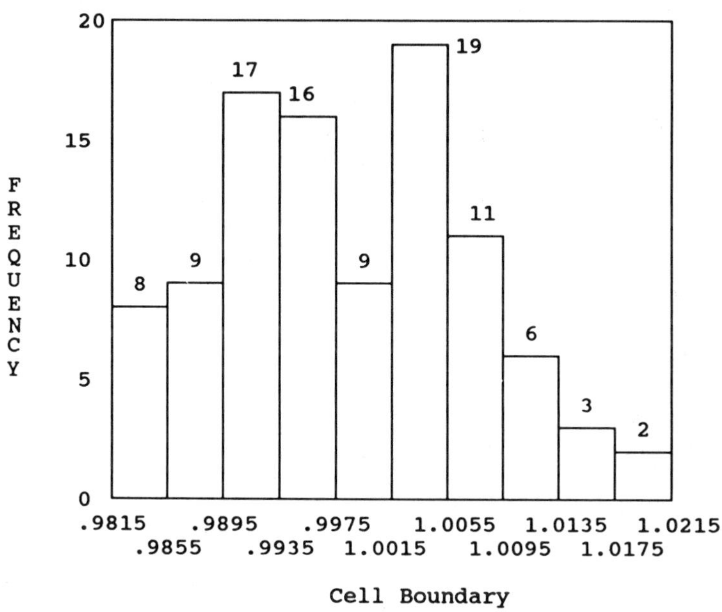

Figure 7.11 Histogram.

tools used to evaluate the dispersion or spread of a process over time.

Usage. Average charts answer the question, "Has a special cause of variation caused the central tendency of this process to change over the time period observed?" Range charts answer the question, "Has a special cause of variation caused the process distribution to become more or less erratic?" Range charts are nearly always used in conjunction with another SPC chart that evaluates the central tendency of the process. Average and range charts can be applied to any continuous variable such as weight or size.

How to Prepare and Analyze Average and Range Charts

1. Determine the subgroup size and sampling frequency. Typically, subgroups of four to six units are sufficient and subgroups of five are most common. The subgroup size affects the sensitivity of the control chart. Smaller subgroups give a control chart that is less sensitive to changes in the process, while larger subgroups are too sensitive to small, unimportant changes. Another factor to consider is the central limit theorem, for distributions that are extremely nonnormal in shape subgroups of at least four are necessary to assure that averages will be approximately normally distributed. Selection of subgroups should be planned to minimize possible variation; this is usually accomplished by selecting consecutive units. Sampling should be frequent enough to detect the effect of special causes while the special cause itself can still be identified. This varies a great deal from process to process, but as a rule you should average about 1 out-of-control point per typical control chart of 25 groups. If you have more than that, increase the sampling frequency. If you have fewer, reduce the sampling frequency. See Montgomery (1986) on using cost models to establish sampling frequency and subgroup size.

2. Collect data from 20 to 25 subgroups; at least 100 individual values are recommended. While the data are being collected, minimize disturbances to the process. If a process change is unavoidable, develop a system for recording changes so that their effect can be determined.

3. Perform the necessary calculations. The flowchart in Figure 7.12 is reprinted from the June 1986 ASQC Statistics Division Newsletter. The form was developed by Barry Griffin of Valparaiso University. Use of the form is quite simple. Each arithmetic instruction in the com-

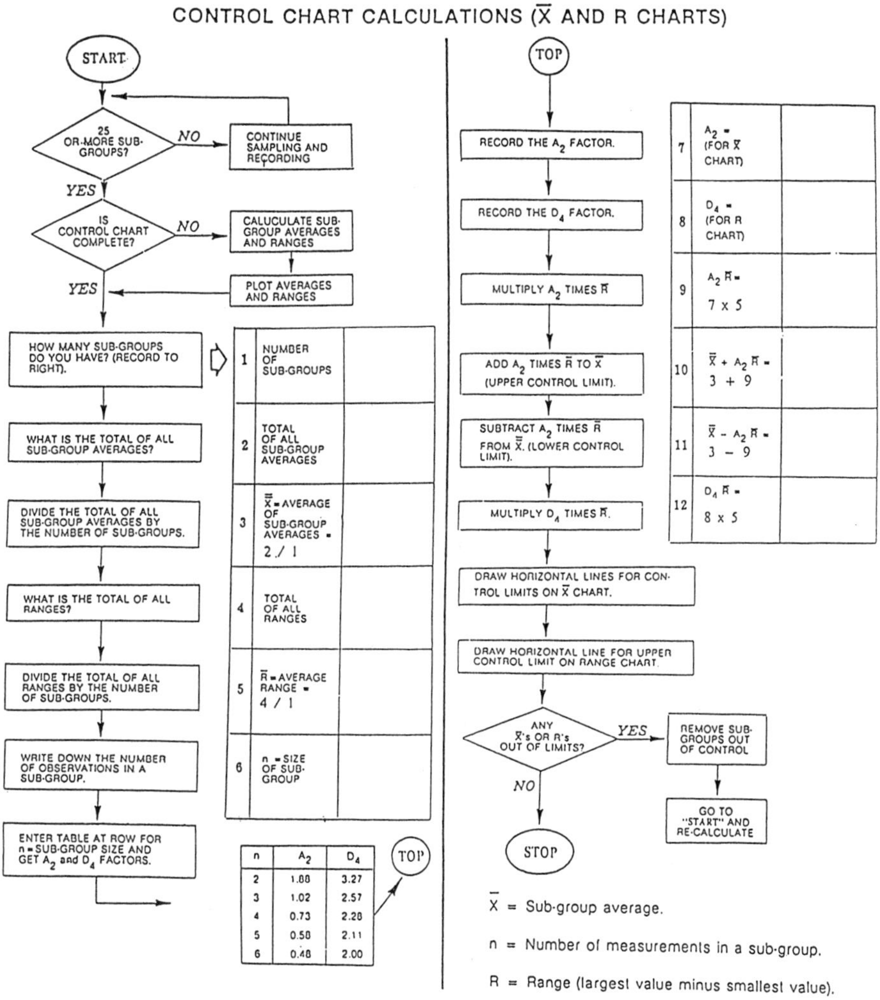

n	A_2	D_4
2	1.88	3.27
3	1.02	2.57
4	0.73	2.28
5	0.58	2.11
6	0.48	2.00

Figure 7.12 Flowchart for X-bar and R chart calculations.

putational sequence is accompanied by a rectangle divided into three segments: a boldface identification number, a description of the data element or instruction, and a blank space for recording the resulting data. Boldface numbers written beneath the description or instruction in the second segment refer you back to the corresponding rectangles. In the descriptive portion of rectangle number 5, for example, **4/1** means that you should divide the data element written in rectangle number 4 by the data element in rectangle 1. By stepping through this simple logic, you will calculate control limits for both *X*-bar and *R* charts by the time you complete rectangle number 12.

4. Interpretation of control charts. The interpretation of control charts for the average and the range involves two considerations: freaks and nonrandom patterns. Either case represents the presence of a special cause of variation. Freaks are detected by comparing each individual subgroup average and range to the control limits computed in rectangles 10, 11, and 12 in the flowchart. The upper control limit for the averages is the value in rectangle 10; the formula is

$$\mathrm{UCL}_{\bar{X}} = \bar{\bar{X}} + A_2\bar{R} \tag{22}$$

The A_2 value is a constant from Table 6 of the Appendix. All control limits are 3σ limits; however, we are using the average range instead of σ as a measure of spread. The A_2 constants account for this, and they also account for the fact that we are plotting averages instead of individual values. Thus, the upper control limit is in reality 3 standard deviations of the mean above the grand average. The situation is analogous for the lower control limit for the averages, which is the value in rectangle 11. The equation for the lower control limit is

$$\mathrm{LCL}_{\bar{X}} = \bar{\bar{X}} - A_2\bar{R} \tag{23}$$

The upper control limit for the ranges is given in rectangle 12; for subgroups of less than 7 the lower control limit for the ranges is 0. The control limits for the range chart are

$$\mathrm{LCL}_R = D_3\bar{R} \tag{24}$$
$$\mathrm{UCL}_R = D_4\bar{R} \tag{25}$$

As before, the D_3 and D_4 values are in Table 6 and are designed so that the control limits are 3σ on either side of the average. In this case we are talking about 3σ of the range. Note that for subgroups smaller than 7, $D_3 = 0$, which gives a lower control limit of zero. Because of this the table included with the flowchart, which only goes up to subgroup size 5, doesn't give D_3 values.

5. Control charts and the normal distribution. Under the assumption that the process distribution is normal, the probability of a single point exceeding a control limit is 2 in 741. This is illustrated in Figure 7.13. The diagram shows that the average or range chart can be thought of as a normal distribution "stretched out in time." With this in mind we can also analyze the *patterns* on the control charts to see if they might be attributed to a special cause of variation. To do this we begin with Figure 7.13 and add two more lines on each side of the grand average, at 1 sigma and 2 sigma, respectively. Since, as shown on the diagram, the control limits are at plus and minus 3 sigma, finding the 1 and 2 sigma lines is as simple as dividing the zone between the grand average and either control limit into thirds. This divides each half of the control chart into three zones. The three zones are labeled A, B, and C in Figure 7.14. Using the guide in Figure 7.14, tests for nonrandom patterns are applied. Remember, the existence of a nonrandom pattern means that a special cause of variation was (or is) probably present. Our X-bar control chart tests are shown in Figure 7.15 (Nelson, 1984). Note that when a point responds to an out-of-control test it is marked with an "x" to make the interpretation of the chart easier. Using this

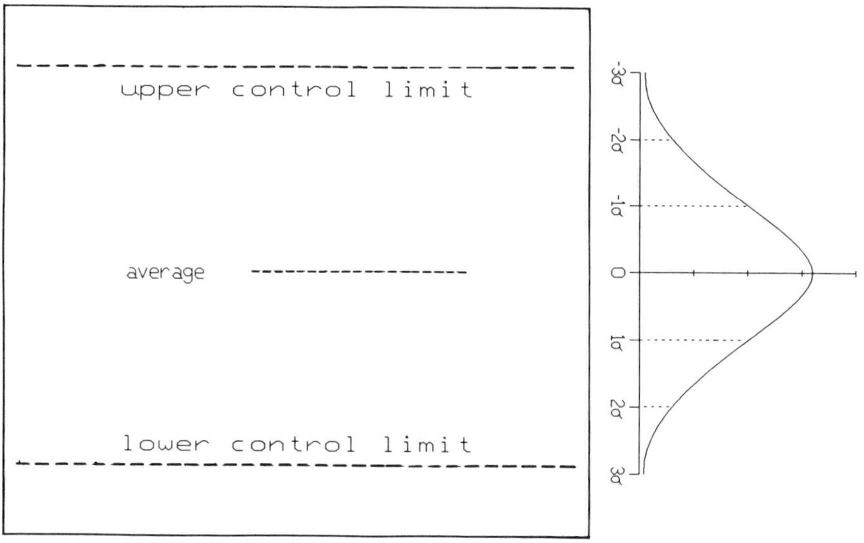

Figure 7.13 How a control chart is related to the normal distribution.

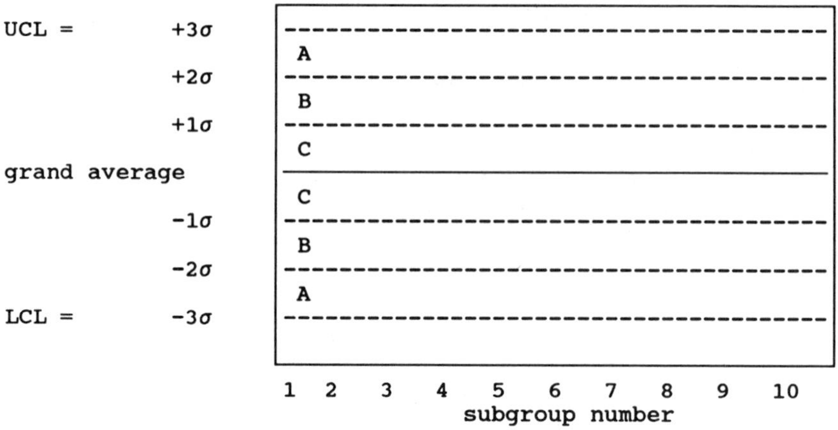

Figure 7.14 Zones on a control chart.

convention, the patterns on the control charts can be used as an aid in troubleshooting.

Example. The table below contains 20 subgroups of 5 observations each. The average and range for each subgroup are computed using formulas (18) and (19).

Row			Data		
1	1.002	0.995	1.000	1.002	1.005
2	1.000	0.997	1.007	0.992	0.995
3	0.997	1.013	1.001	0.985	1.002
4	0.990	1.008	1.005	0.994	1.012
5	0.992	1.012	1.005	0.985	1.006
6	1.000	1.002	1.006	1.007	0.993
7	0.984	0.994	0.998	1.006	1.002
8	0.987	0.994	1.002	0.997	1.008
9	0.992	0.988	1.015	0.987	1.006
10	0.994	0.990	0.991	1.002	0.988
11	1.007	1.008	0.990	1.001	0.999
12	0.995	0.989	0.982	0.995	1.002
13	0.987	1.004	0.992	1.002	0.992
14	0.991	1.001	0.996	0.997	0.984
15	1.004	0.993	1.003	0.992	1.010

Row			Data		
16	1.004	1.010	0.984	0.997	1.008
17	0.990	1.021	0.995	0.987	0.989
18	1.003	0.992	0.992	0.990	1.014
19	1.000	0.985	1.019	1.002	0.986
20	0.996	0.984	1.005	1.016	1.012

Using these data, we complete the form as shown in Figure 7.16. Note that when completing the form for rectangles 10 and 11 we carried two more decimal places than the raw data contain. It is recommended that the average be rounded to one decimal place more than the raw data. The range control limit in rectangle 12 has one more decimal place than the raw data. The above conventions eliminate the ambiguity that might result if a sample average or range fell exactly on the control limit. The completed control charts are shown in Figure 7.17. There are some minor differences between the control chart limits and the figure 17.16 worksheet. These are due to roundoff errors.

Control Charts for Proportion Defective (*p* Charts)

Definition. *p* charts are statistical tools used to evaluate whether the proportion nonconforming produced by a process demonstrates a state of statistical control. The determinations are made through comparison of the values of some statistical measures for an ordered series of samples, or subgroups, with control limits.

Usage. *p* charts can be applied to any variable where the appropriate performance measure is a unit count. These charts answer the question, "Has a special cause of variation caused this process to produce an abnormally large or small number of defective units over the time period observed?"

How to Prepare and Analyze p *Charts*

1. Determine the sampling frequency. Sampling should be frequent enough to detect the effect of special causes while the special cause itself can still be identified. This varies a great deal from process to process, but as a rule you should average about 1 out-of-control point per typical control chart of 25 groups. If you have more than that, increase the sampling frequency. If you have fewer, reduce the sampling frequency.

2. Collect data from 25 to 30 subgroups. While the data are being

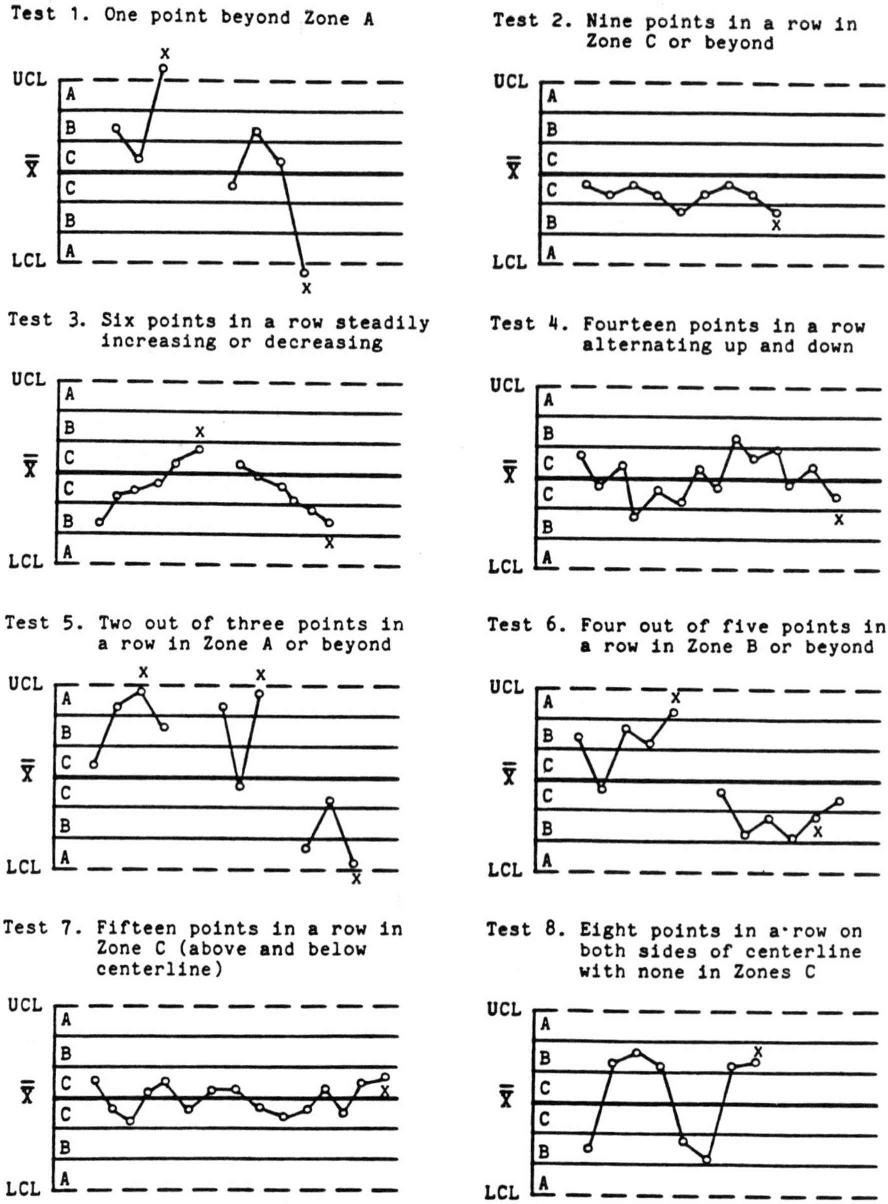

Figure 7.15 Tests for out-of-control patterns on control charts. (From Tests for Out of Control Patterns on Control Charts. Copyright 1984, American Society for Quality Control, Inc., Milwaukee, Wisconsin. Reprinted by permission from American Society for Quality Control.)

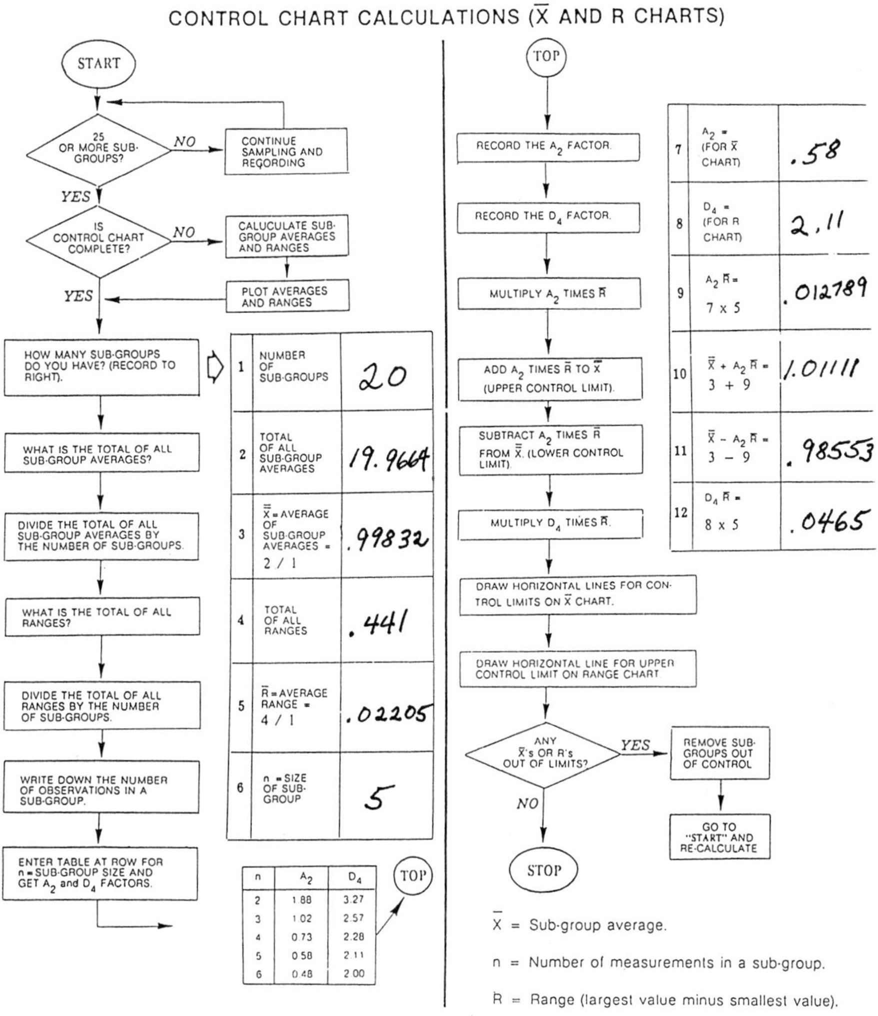

Figure 7.16 Example of completed *X*bar and *R* chart work sheet.

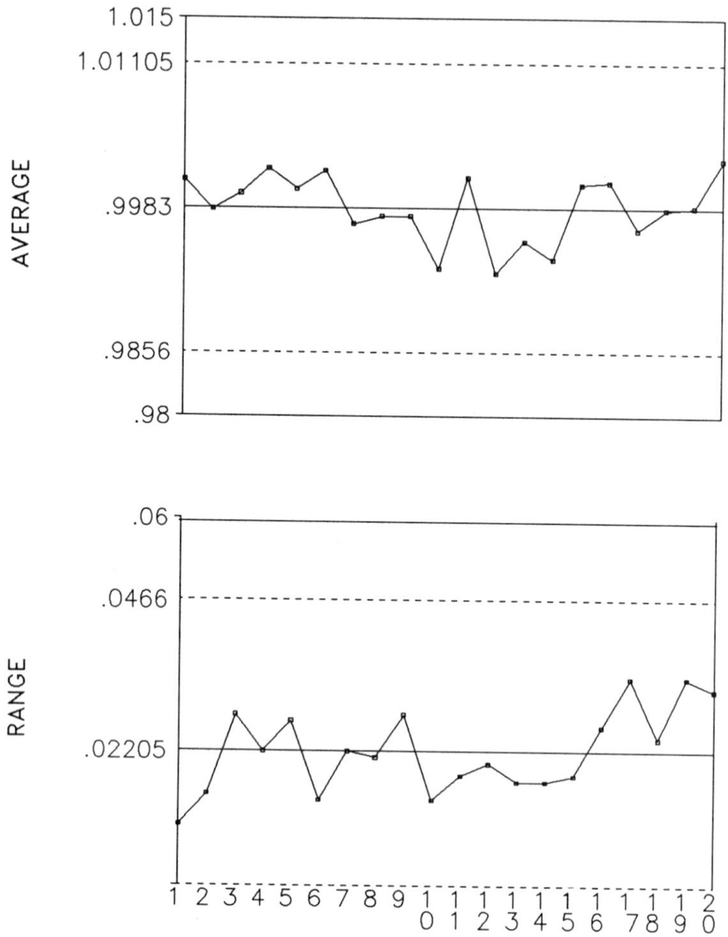

Figure 7.17 Example of completed *X*-bar and *R* Charts.

collected, minimize disturbances to the process. If a process change is unavoidable, develop a system for recording changes so that their effect can be determined.

3. Establish the control limits. Perform the necessary calculations. The worksheet (Figure 7.18) will be helpful in calculating the control limits. Note that the control limits vary as the sample size varies. Thus,

for *p* charts there are separate control limits for every subgroup. Also note that control limits should be based on data taken when the process was in control. this means that you should identify the special causes of out-of-control subgroups and drop these groups from your calculations. The worksheet solves three equations: the average proportion

1. Collect information on 25–30 subgroups.
2. Record the size of each subgroup in column 2 of the worksheet on the next page.
3. Record the number defective in each subgroup in column 3 of the worksheet.
4. Compute the subgroup proportion defective, p, by dividing the subgroup number defective (column 3) by the subgroup size (column 2). Record the result in column 3. NOTE: When using the worksheet, any number in parentheses ( ) refers to a column number. e.g. (3)÷(2) means to divide the value in column 3 by the value in column 2.
5. Compute the square root of each subgroup size and record the result in column 5. NOTE: The notation sqrt{x} means to take the square root of x, the value in the braces { }.
6. Add the sample sizes for all of the subgroups. Record the result at the bottom of column 2 and in the box below labeled "n = sum of all subgroups."
7. Add the sample defectives for all of the subgroups. Record the result at the bottom of column 2 and in the box below labeled "d = sum of all defectives."
8. Complete the calculations for all of the remaining boxes on this page. The numbers that go in each box are always computed in a previous step.

```
n = sum of all subgroups                 [      ]

d = sum of all defectives                [      ]

pbar = d÷n = [      ]  ÷  [      ]   =   [      ]

qbar = 1 - pbar  = 1  -  [      ]   =   [      ]

A = pbar x qbar  =  [      ]  x  [      ]  =  [      ]

B = sqrt { A }   = sqrt { [      ] }=  [      ]

C = 3 x B  = 3 x  [      ]  =  [      ]
```

Figure 7.18 *p* Chart Worksheet Instructions.

(1)	(2)	(3)	(4)	(5)	(6)	(7)	(8)
Subgroup Number	Subgroup Size	Subgroup Defectives	p = (3)÷(2)	sqrt of (2)	C ÷ (5)	LCL = pbar-(6)	UCL = pbar+(6)
1							
2							
3							
4							
5							
6							
7							
8							
9							
10							
11							
12							
13							
14							
15							
16							
17							
18							
19							
20							
21							
22							
23							
24							
25							
totals							

Figure 7.18 *(Continued)*

defective, the upper control limit for the fraction defective, and the lower control limit for the fraction defective. The upper and lower control limits must be solved for each subgroup since the standard deviation of a proportion varies as the sample size varies. The equations being solved by the worksheet are

$$\bar{p} = \frac{\text{total number defective}}{\text{total number inspected}} \tag{26}$$

$$\text{LCL}_P = \bar{p} - 3\sqrt{\frac{\bar{p}(1-\bar{p})}{n_i}} \tag{27}$$

$$\text{UCL}_P = \bar{p} + 3\sqrt{\frac{\bar{p}(1-\bar{p})}{n_i}} \tag{28}$$

where n_i is the size of each individual subgroup. As with average and range charts, the control limits are set symmetrically 3σ on either side of the average. However, with p charts the lower control limit sometimes falls below zero; in these cases it is simply set equal to zero. If the upper control limit is greater than one it is set equal to one.

4. Analysis of p charts. As with all control charts, a special cause is probably present if there are any points beyond either the upper or the lower control limit. Analysis of p chart patterns between the control limits is extremely complicated because the sample size varies and the distribution of the number defective varies with the sample size. Because of this, pattern analysis as described for X-bar charts is not done for p charts. A method known as a stabilized p chart converts all of the proportions to standard deviation units and *does* allow constant control limits and pattern analysis. The reader is referred to Duncan (1974) for additional information on stabilized p charts.

Example. Bob, our friendly neighborhood produce manager, has kept track of the number of bruised peaches in recent shipments. The following data were obtained by opening randomly selected crates from each shipment and counting the number of bruised peaches. There are 250 peaches per crate. Normally Bob samples only one crate per shipment. However, when his part-time helper is available Bob samples two crates.

Shipment	Crates checked	Peaches checked	Bruised peaches
1	1	250	47
2	1	250	42
3	1	250	55
4	1	250	51
5	1	250	46
6	1	250	61
7	1	250	39
8	1	250	44
9	1	250	41
10	1	250	51
11	2	500	88
12	2	500	101
13	2	500	101
14	1	250	40
15	1	250	48
16	1	250	47
17	1	250	50
18	1	250	48
19	1	250	57
20	1	250	45
21	1	250	43
22	2	500	105
23	2	500	98
24	2	500	100
25	2	500	96

```
n = sum of all subgroups          [ 8000 ]

d = sum of all defectives         [ 1544 ]

pbar = d÷n = [ 1544 ] ÷ [ 8000 ] = [ .193 ]

qbar = 1 - pbar = 1 - [ .193 ] = [ .807 ]

A = pbar x qbar = [ .193 ] x [ .807 ] = [ .155751 ]

B = sqrt { A } = sqrt { [ .155751 ] }= [ .39465 ]

C = 3 x B = 3 x [ .39465 ] = [ 1.1840 ]
```

Figure 7.19 Example of completed *p*-chart worksheet.

(1)	(2)	(3)	(4)	(5)	(6)	(7)	(8)
Subgroup Number	Subgroup Size	Subgroup Defectives	p = (3)÷(2)	sqrt of (2)	C ÷ (5)	LCL = pbar-(6)	UCL = pbar+(6)
1	250	47	.188	15.8114	.07488	.1181	.2679
2	250	42	.168	"	"	"	"
3	250	55	.220	"	"	"	"
4	250	51	.204	"	"	"	"
5	250	46	.184	"	"	"	"
6	250	61	.244	"	"	"	"
7	250	39	.156	"	"	"	"
8	250	44	.176	"	"	"	"
9	250	41	.164	"	"	"	"
10	250	51	.204	"	"	"	"
11	500	88	.176	22.3607	.05295	.1400	.2460
12	500	101	.202	"	"	"	"
13	500	101	.202	"	"	"	"
14	250	40	.160	15.8114	.07488	.1181	.2679
15	250	48	.192	"	"	"	"
16	250	47	.188	"	"	"	"
17	250	50	.200	"	"	"	"
18	250	48	.192	"	"	"	"
19	250	57	.228	"	"	"	"
20	250	45	.180	"	"	"	"
21	250	43	.172	"	"	"	"
22	500	105	.210	22.3607	.05295	.1400	.2460
23	500	98	.196	"	"	"	"
24	500	100	.200	"	"	"	"
25	500	96	.192	"	"	"	"
totals	8000	1544					

Figure 7.19 (*Continued*)

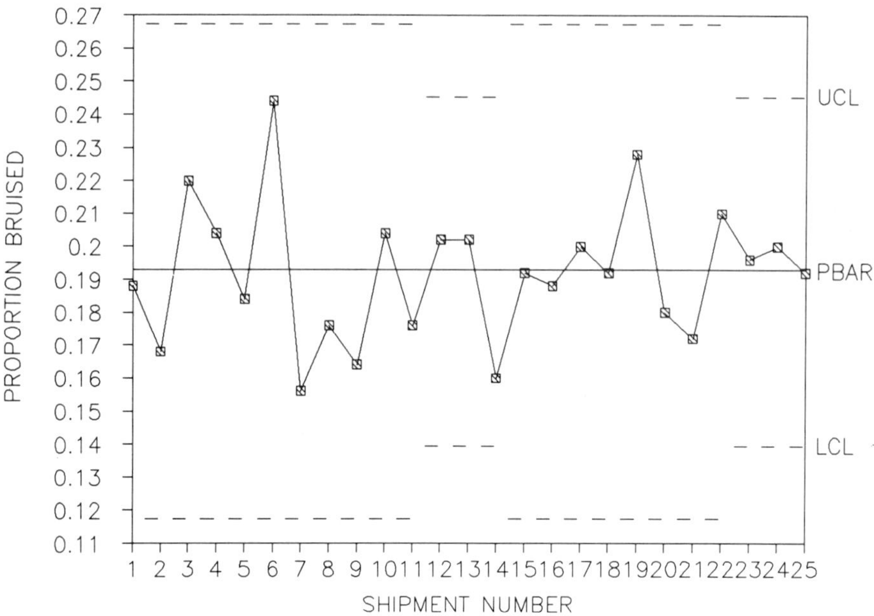

Figure 7.20 Example of completed *p* chart.

5. Pointers for real-world use. Determine if "moving control limits" are really necessary. It may be possible to use the average sample size (total number inspected divided by number of subgroups) to calculate control limits. For instance, in the example above the sample size doubled from 250 to 500 peaches but the control limits hardly changed at all. The following table illustrates the different control limits based on 250 peaches, 500 peaches, and the average sample size, which is 8000/25 = 320 peaches.

Sample size	Lower control limit	Upper control limit
250	.1181	.2679
500	.1400	.2460
320	.1268	.2592

Notice that our conclusions regarding process performance are the same when using the average sample size as when using the exact sample sizes. This is usually the case if the variation in sample size isn't too great. There are many rules of thumb, but most of them are extremely conservative. The best way to evaluate limits based on the average sample size is to check it out the way we did here.

Other SPC Techniques

A wide variety of other SPC techniques exist. A complete discussion of these techniques is beyond the scope of this book. See Montgomery (1986) for some examples.

Process Capability Analysis

Definition. Process capability analysis is a two-stage process that involves (1) bringing a process into a state of statistical control for a reasonable period of time and (2) comparing the long-term process performance to management or engineering requirements.

Usage. Process capability analysis can be done with either attribute data or continuous data *if and only if the process is in statistical control* and has been for a reasonable period of time. Two variations on traditional process capability analysis allow use of the statistical approach with limited data: the process potential study, where only 30 consecutive units are checked (Ford Motor Company), and the mini-capability study, where only 10 consecutive units are checked (General Motors). Application of process capability methods to processes that are not in statistical control results in unreliable estimates of process capability and should never be done. Figure 7.21 illustrates these concepts.

*How to Perform Process Capability Studies, a 10-Step Plan**

1. Select a candidate for the study. This step should be institutionalized. A goal of any organization should be ongoing process improvement. However, because a company has only a limited resource base and can't solve all problems simultaneously, it must set priorities

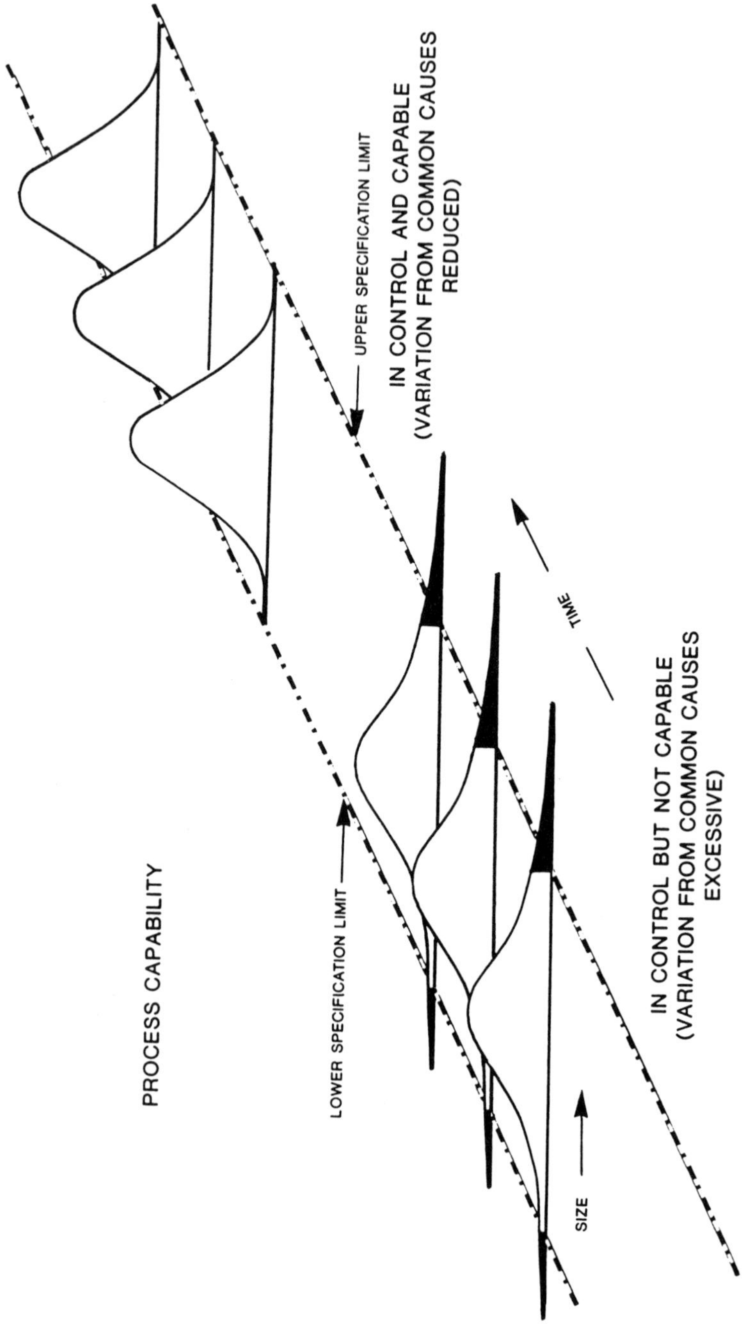
PROCESS CAPABILITY
UPPER SPECIFICATION LIMIT
LOWER SPECIFICATION LIMIT
IN CONTROL AND CAPABLE
(VARIATION FROM COMMON CAUSES
REDUCED)
TIME
SIZE
IN CONTROL BUT NOT CAPABLE
(VARIATION FROM COMMON CAUSES
EXCESSIVE)

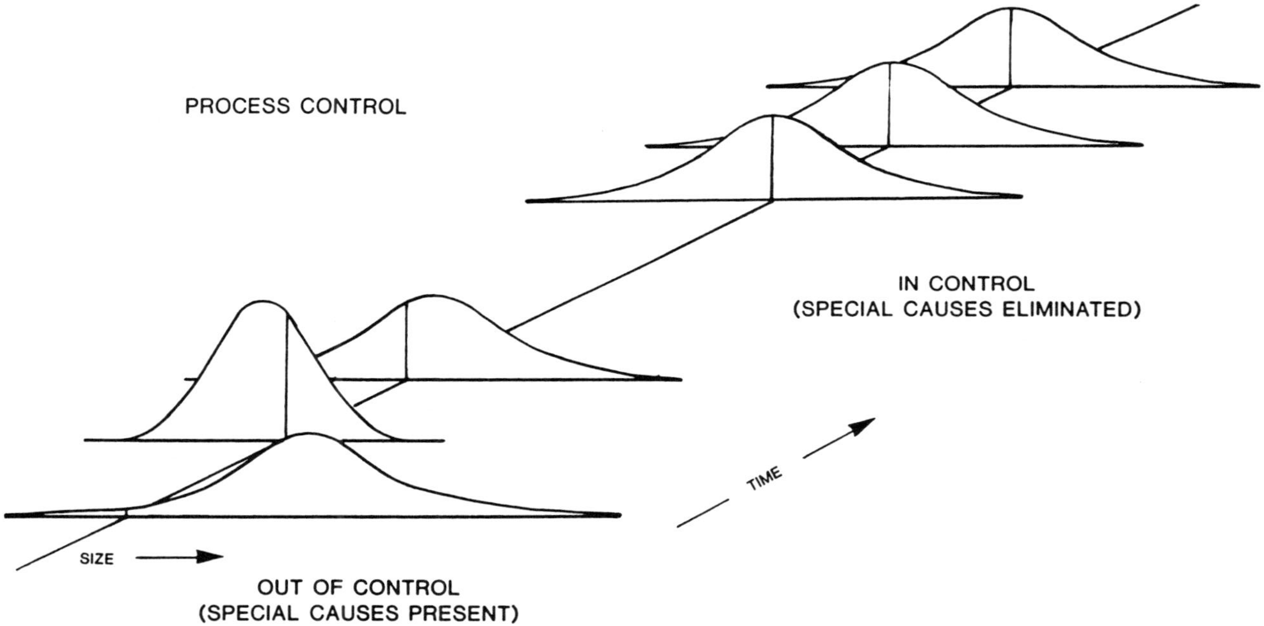

Figure 7.21 Process control concepts illustrated. (From *Continuing Process Control and Process Capability Improvement*, p. 4a. Copyright 1983. Used by permission of the publisher, Ford Motor Company, Dearborn, Michigan.)

for its efforts. The tools for this include Pareto analysis and fishbone diagrams.

2. Define the process. It is all too easy to slip into the trap of solving the wrong problem. Once the candidate area has been selected in step 1, define the scope of the study. A process is a unique combination of machines, tools, methods, and personnel engaged in production. Each element of the process should be identified at this stage. This is not a trivial exercise. The input of a large number of people may be required. There are likely to be a number of conflicting opinions about what the process actually involves. These differences must be resolved into a consensus. Finally, when the consensus has been achieved, a flowchart of the process must be prepared.

3. Procure resources. Process capability studies disrupt production and require significant expenditures of both material and human resources. Since this is a project of major importance, it should be managed as such. All of the usual project management techniques should be brought to bear. This includes planning, scheduling, and management status reporting.

4. Evaluate the measurement system. Using the techniques described in Chapter 11, evaluate the measurement system's ability to do the job. Again, be prepared to spend the time necessary to get a valid means of measuring the process before going ahead.

5. Provide a control system. The purpose of the control system is twofold: (1) isolate and control as many important variables as possible, and (2) provide a mechanism for tracking variables that cannot be adequately controlled. The object of the capability analysis is to determine what the process can do if it is operated as it should be. This means that such obvious sources of potential variation as operators and vendors will be controlled while the study is conducted. In other words, a single well-trained operator will be used and the material will be from a single vendor.

There are usually some variables that are important but are not controllable. One example is the ambient environment, such as temperature, barometric pressure, or humidity. Certain process variables may degrade as part of the normal operation; for example, tools wear and chemicals are used. These variables should still be tracked using logsheets and similar tools.

6. Select an SPC method to use for the analysis. The SPC method will depend on the decisions made up to this point. If the performance

measure is an attribute, one of the attribute charts will be used. Variables charts will be used for process performance measures assessed on a continuous scale. Also considered will be the skill level of the personnel involved, need for sensitivity, and other resources required to collect, record, and analyze the data.

7. Gather and analyze the data. This entire chapter is devoted to techniques to help you do this. It is usually advisable to have at least two people go over the data analysis to catch inadvertent erors in transcribing data or performing the analysis.

8. Track down and remove special causes of variation. A special cause of variation may be obvious, or it may take months of painstaking investigation to find it. The effect of the special cause may be good or bad. Removing a special cause that has a bad effect usually involves eliminating the cause itself. For example, if poorly trained operators are causing variability, the special cause is the training system (not the operator) and it is eliminated by developing an improved training system or a process that requires less training. However, the removal of a beneficial special cause may actually involve incorporating the special cause into the normal operating procedure. For example, if it is discovered that materials with a particular chemistry produce a better product, the special cause is the newly discovered material and it can be made a common cause simply by changing the specification to assure that the new chemistry is always used.

9. Estimate process capability. One point cannot be overemphasized: the process capability cannot be estimated until a state of statistical control has been achieved. After this stage has been reached, the methods described in the next section may be used. After the numerical estimate of process capability has been arrived at, it must be compared to management's goals for the process, or it can be used as an input into economic models. Deming (1982) describes a simple economical model that can be used to determine if the output from a process should be sorted 100% or shipped as is.

10. Establish a quality improvement and control plan for the process. Once a stable process state has been attained steps should be taken to maintain it and, hopefully, improve on it. Statistical process control is one means of doing this. Others include employee involvement programs, quality circles, and quality of work life programs. Far more important than the particular approach taken is a company environment that makes continuous improvement a normal part of the daily routine.

Statistical Analysis of Process Capability Data

Control Chart Method: Attributes Data

1. Collect samples from 25 or more subgroups of consecutively produced units. Use the guidelines presented in the appropriate control chart section to determine the subgroup size. Follow the guidelines presented in the section on process capability studies for setting up and conducting the study.

2. Plot the results on the appropriate control chart (e.g., p chart). If all groups are in statistical control, go to the next step. Otherwise identify the special cause of variation, take action to eliminate it, and repeat step 1.

3. Using the control limits from the previous step (called operation control limits), put the control chart to use for a period of time. Once you are satisfied that sufficient time has passed for most special causes to have been identified and eliminated, as verified by the control charts, go to the next step.

4. The process capability is estimated as the control chart *centerline.* The centerline on attribute charts is the long-term expected quality level of the process (e.g., the average proportion defective). This is the level created by the common causes of process variation.

5. If the process capability doesn't meet management requirements, take immediate action to modify the process for the better. Whether it meets requirements or not, always be on the lookout for possible process improvements. The control charts will provide verification of improvement.

Control Chart Method: Variables Data

1. Collect samples from 25 or more subgroups of consecutively produced units, following the 10-step plan described above.

2. Plot the results on the appropriate control chart (e.g., X-bar and R chart). If all groups are in statistical control, go to the next step. Otherwise identify the special cause of variation, take action to eliminate it, and repeat step 1.

3. Using the control limits from the previous step (called operation control limits), put the control chart to use for a period of time. Once you are satisfied that sufficient time has passed for most special causes to have been identified and eliminated, as verified by the control charts, go to the next step.

4. The process capability is estimated from the process average and standard deviation. The standard deviation is computed based on the average range, in the case of individual charts the average difference is used. This is the level of variability created by the common causes of process. The formula is

$$\hat{\sigma} = \frac{\bar{R}}{d_2} \qquad \text{[Equation (20) repeated]}$$

5. The value d_2 was discussed earlier. It is a constant that is obtained from Table 6 in the Appendix.

6. Only now can the process be compared to engineering requirements. One way of doing this is by calculating "capability indexes." Following are several popular capability indexes and their interpretation.

Process Capability Indexes

$$C_p = \frac{\text{engineering tolerance}}{6\hat{\sigma}} \tag{29}$$

$$C_R = 100 \times \frac{6\hat{\sigma}}{\text{engineering tolerance}} \tag{30}$$

$$Z_U = \frac{\text{upper specification} - \text{process average}}{\hat{\sigma}} \tag{31}$$

$$Z_L = \frac{\text{process average} - \text{lower specification}}{\hat{\sigma}} \tag{32}$$

$$Z_{\min} = \text{Minimum}\{Z_L, Z_U\} \tag{33}$$

$$C_{pk} = \frac{Z_{\min}}{3} \tag{34}$$

These capability indexes are listed below with interpretations and comments.

C_p This is the most common capability index. It simply makes a direct comparison of the process to the engineering requirements. Assuming the process distribution is normal and the process average is exactly centered between the engineering requirements, a C_p index of 1 would give a "capable process." However, to allow a bit of room for process drift, the generally accepted minimum value for C_p is 1.33. This situation is shown in Figure 7.22. In general, the larger C_p is, the better. The C_p index has two major shortcomings: it can't be used unless there are both upper and lower specifications, and it does not account for process centering. If the process average is not exactly centered relative to the engineering requirements, the C_p index will give misleading results.

C_R The C_R index is exactly equivalent to the C_p index. It simply makes a direct comparison of the process to the engineering requirements. Assuming the process distribution is normal and the process average is exactly centered between the engineering requirements, a C_R index of 100 would give a "capable process." However, to allow a bit of room for process drift, the generally accepted maximum value for C_R is 75. In general, the smaller C_R is, the better. The C_R index suffers from the same shortcomings as the C_p index.

Z_U The Z_U index measures the process location (central tendency) relative to its standard deviation and the upper requirement. If the distribution is normal, the value of Z_U can be used to determine the percentage above the upper requirement by using

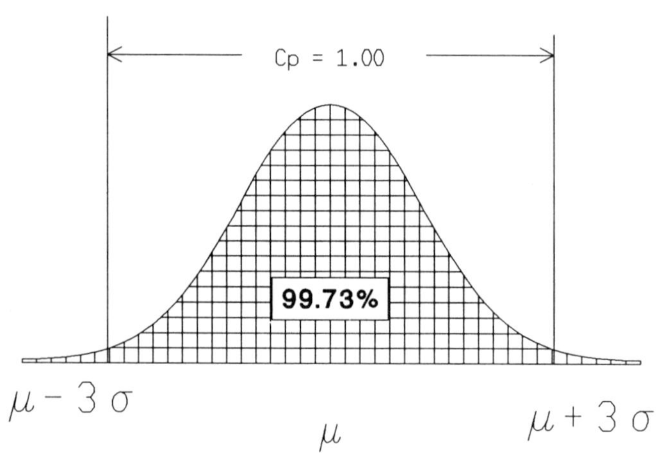

Figure 7.22 Illustration of a capable process.

Table 1 in Appendix. The method is the same as described in the previous chapter using the Z transformation; simply use Z_U instead of Z. In general, the bigger Z_U is, the better. A value of at least +3 is required to assure that 0.1% or less defective will be produced. A value of +4 is generally desired to allow some room for process drift.

Z_L The Z_L index measures the process location relative to its standard deviation and the lower requirement. If the distribution is normal, the value of Z_L can be used to determine the percentage above the upper requirement by using Table 1 in the Appendix. The method is the same as described in the previous chapter using the Z transformation, except that you use $-Z_L$ instead of Z. In general the bigger Z_L is, the better. A value of at least +3 is required to assure that 0.1% or less defective will be produced. A value of +4 is generally desired to allow some room for process drift.

Z_{min} The value of Z_{min} is simply the smaller of the Z_L and Z_U values. It can be considered an intermediate result to be used in computing C_{pk}.

C_{pk} The value of C_{pk} is simply Z_{min} divided by 3. Since the smallest value represents the nearest specification, the value of C_{pk} tells you if the process is truly capable of meeting requirements. A C_{pk} of at least +1 is required, and +1.33 is preferred. Note that C_{pk} is closely related to C_p; the difference between C_{pk} and C_p represents the potential gain to be had from centering the process.

Example of Capability Analysis Using Normally Distributed Variables Data. Assume we have conducted a capability analysis using X-bar and R charts with subgroups of 5. Also assume that we found the process to be in statistical control with a grand average of 0.9983 and an average range of 0.02205. From the table of d_2 values (Appendix Table 6) we find d_2 is 2.326 for subgroups of 5. Thus, using equation (20).

$$\hat{\sigma} = \frac{0.2205}{2.326} = 0.00948$$

Before we can analyze process capability, we must know what the requirements are. For this process the requirements are a lower specification of 0.980 and an upper specification of 1.020 (1.000 plus or minus 0.020). With this information, plus the knowledge that the process performance has been in statistical control, we can compute the capability indexes for this process.

$$C_p = \frac{1.020 - 0.9800}{6 \times 0.00948} = \frac{0.04}{0.05688} = 0.703$$

$$C_R = 100 \times \frac{0.05688}{0.04} = 142.20\%$$

$$Z_U = \frac{1.020 - 0.99832}{0.00948} = \frac{0.02168}{0.00948} = +2.3$$

$$Z_L = \frac{0.99832 - 0.980}{0.00948} = \frac{0.01832}{0.00948} = +1.9$$

$$Z_{\min} = \text{Minimum}\{1.9, 2.3\} = 1.9$$

$$C_{pk} = \frac{1.9}{3} = 0.63$$

OK, so what does all of this mean? The answer is found by interpreting the capability indexes. Let's look at them one at a time.

Index	Value	Interpretation
C_p	0.703	Since the minimum acceptable value for this index is 1, the result 0.703 indicates that this process cannot meet the requirements. Furthermore, since the C_p index doesn't consider process centering, we know that the process can't be made acceptable by merely adjusting the process closer to the center of the requirements. Thus we can expect the Z_L, Z_U, and $Z_{\min}$ values to be unacceptable too.
C_R	142.2%	This value always gives the same conclusions as the C_p index. The number itself means that the "natural tolerance" of the process uses 142.2% of the engineering requirement.
Z_U	+2.3	We desire a Z_U of at least +3, so this value is unacceptable, as we predicted from the C_p index. We can use Z_U to estimate the percentage of production that will exceed the upper specification. Referring to the interpretation of Z_U above, we find that approximately 1.1% will be oversized.

Index	Value	Interpretation
Z_L	+1.9	We desire a Z_L of at least +3, so this value is unacceptable, as we predicted from the C_p index. We can use Z_L to estimate the percentage of production that will be below the lower specification. Referring to the interpretation above, we find that approximately 2.9% will be undersized and we estimate a total reject rate of 4.0%. By subtracting this from 100% we get the projected yield of 96.0%.
$Z_{\min}$	+1.9	Obviously, this value will not be acceptable.
C_{pk}	0.63	The value of C_{pk} is only slightly smaller than that of C_p. This indicates that we will not gain much by centering the process. The actual amount we would gain can be calculated by "assuming" the process is exactly centered at 1.000 and recalculating $Z_{\min}$; if we do this we find the predicted total reject rate drops from 4.0 to 3.6%.

SUMMARY

This chapter introduced the concept of statistical control. Statistical process control was defined. Variation as a fact of life was discussed, along with ways of characterizing variation in a process. The central limit theorem was presented (without proof) and its importance in SPC was explained. The differences between prevention-based control systems and systems based on detection were presented. The SPC techniques of Pareto analysis, cause-and-effect diagrams, histograms, X-bar and range charts, p charts, and process capability analysis were introduced along with worked-out examples.

This chapter did not cover the relationships among the various SPC techniques, and the application of the SPC techniques to troubleshooting was not discussed (e.g., the patterns that appear on the histograms and control charts can often be related to causes listed on cause-and-effect diagrams). Omitted were various additional control charts for both attributes and variables data. Also omitted were a number of alternatives to traditional control charting for process control, such as analysis of variance (ANOVA) and chi-square tests. Finally, process capability analysis of nonnormally distributed data was not discussed.

RECOMMENDED READING LIST

48, 58–59, 61–62.

8

Acceptance Sampling

This chapter presents procedures for acceptance sampling. Acceptance sampling is a popular quality control technique that is applied to discrete lots or batches of product. The term *lot* refers to collection of physical units; the term *batch* is usually applied to chemical materials. The lot or batch is typically presented to the inspection department by either a supplier or a production department. The inspection department then inspects a sample from the lot or batch and, based on the results of the inspection, determines the acceptability of the lot or batch.

Acceptance sampling schemes generally consist of three elements:

1. The sampling plan. How many units should be inspected? What is the acceptance criterion?
2. The action to be taken on the current lot or batch. Actions include accept, sort, scrap, rework, downgrade, return to vendor, etc.
3. The action to be taken in the future. Future actions include such options as switching to reduced or tightened sampling, switching to 100% inspection, or shutting down the process.

CONTINUOUS SAMPLING PLANS

Acceptance sampling can be applied in a variety of situations. When it is applied to a process in continuous production, it is possible to design an acceptance sampling scheme that does not require assembling a discrete lot or batch. Instead, these plans call for the inspection of some fraction of units selected at random as the process is run. They mix 100% inspection with sampling in such a way that the consumer is guaranteed some maximum fraction nonconforming in the outgoing product. Such plans are called continuous sampling plans; for a complete discussion see Schilling (1982). In my opinon, if a process is in continuous production it is generally better to use a control chart for process control than to apply a continuous sampling plan.

PROCEDURES

The remainder of this chapter consists of the essential elements of acceptance sampling presented in procedure format. The procedures describe the most commonly used acceptance sampling schemes and important concepts related to the subject of acceptance sampling. Acceptance sampling is discussed in depth in Schilling (1982).

Terminology and Basic Concepts

Definition of Acceptance Sampling. Sampling inspection in which decisions are made to accept or not accept product or service; also, the methodology that deal with procedures by which decisions to accept or not accept are based on the results of the inspection of samples.

Acceptable Quality Level (AQL). The maximum percentage or proportion of variant units in a lot or batch that, for the purposes of acceptance sampling, can be considered satisfactory as a process average.

Limiting Quality Level (LQL). The percentage or proportion of variant units in a batch or lot for which, for the purposes of acceptance sampling, the consumer wishes the probability of acceptance to be restricted to a specified low value. Also called the lot tolerance percentage defective (LTPD), lot tolerance fraction defective (LTFD), or reject quality level (RQL). The ASQC defines limiting quality level as the preferred term.

Indifference Quality Level (IQL). A quality level that a particular sampling plan will accept 50% of the time.

Producer's Risk. The probability that an acceptable lot, that is, a lot of AQL or better quality, will be rejected.

Consumer's risk. The probability that a lot of rejectable quality will be accepted by a particular sampling plan.

Operating Characteristics (OC) Curve. (1) For isolated or unique lots or a lot from an isolated sequence: a curve showing, for a given sampling plan, the probability of accepting the lot as a funtion of the lot quality (type A). (2) For a continuous stream of lots: a curve showing, for a given sampling plan, the probability of accepting a lot as a function of the process average quality (type B). Figure 8.1 illustrates an OC curve. A perfect OC curve will accept all lots of acceptable quality and reject all other lots. The shape of the perfect curve would be a "Z," indicating the perfect discrimination of the plan.

Average Outgoing Quality (AOQ). The quality level that results when a particular sampling *scheme* (a sampling plan combined with repair or replacement of defectives in the sample and/or lot) is applied to a series of lots from a process.

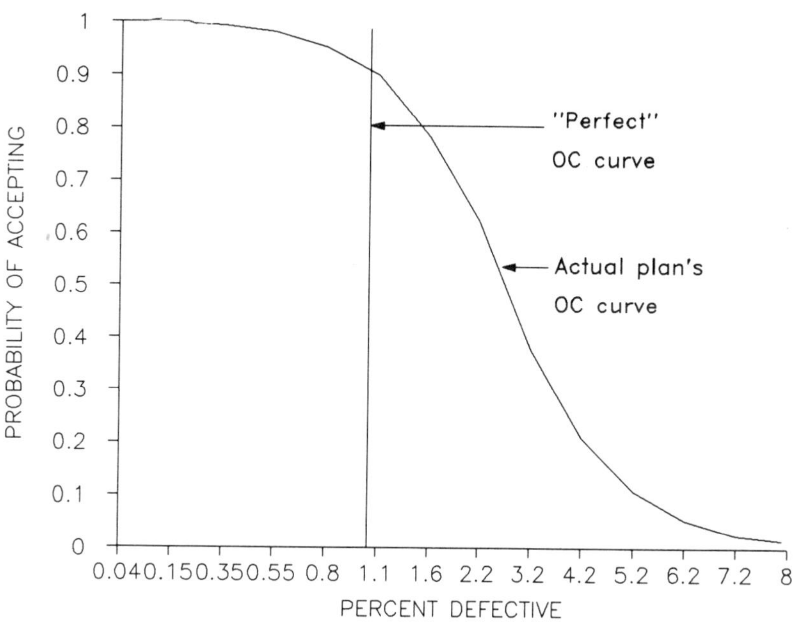

Figure 8.1 Example of operating characteristics curve.

Average Outgoing Quality Limit (AOQL). For a given sampling plan and repair or replacement scheme, the maximum outgoing defective rate for any incoming defective rate. Figure 8.2 illustrates the AOQ and AOQL concepts.

Average Total Inspected (ATI). The average number of units inspected per lot based on the sample size for accepted lots and all inspected units in nonaccepted lots.

Acceptance Number (c). The maximum number of variants or variant units in the sample that will permit acceptance of the inspected lot or batch.

Reject Number. The minimum number of variants or variant units in the sample that will cause the lot to be designated as not acceptable.

Single Sampling Plan. A sampling plan where the acceptance decision is reached after the inspection of a single sample of n items. Single sampling plans work as follows:

Sample size $= n$
Accept number $= c$
Reject number $= c + 1$

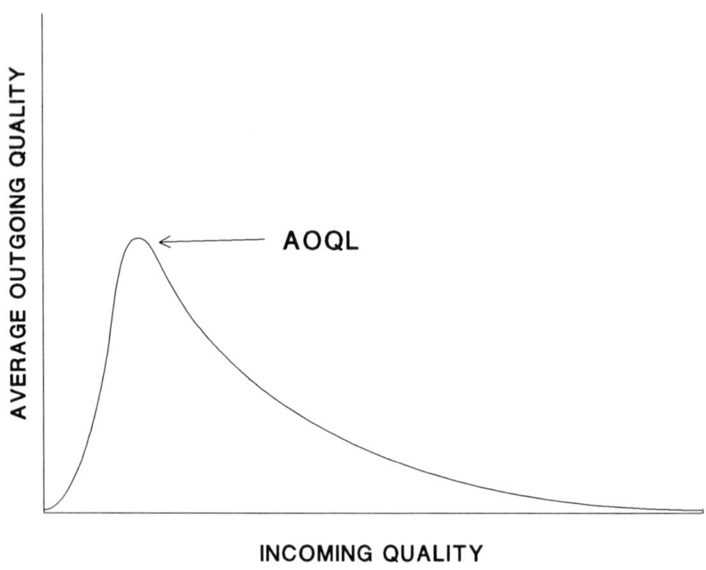

Figure 8.2 Average outgoing quality curve.

For example, if $n = 50$ and $c = 0$ we will inspect a sample of 50 units and accept the lot if there are no defective units in the sample. If there are one or more defectives in the sample, the lot will be rejected.

Double Sampling Plan. Sampling plan where the acceptance decision may not be reached until two samples have been inspected. The samples may or may not be the same size. Double sampling plans work as follows:

Sample number	Sample size	Accept no.	Reject no.
1	n_1	c_1	r_1
2	n_2	c_2	$c_2 + 1$

Note that for the first sample r_1 is at least 2 greater than c_1. An example of a double sampling plan is:

Sample number	Sample size	Accept no.	Reject no.
1	20	1	3
2	25	2	3

This double sampling plan works as follows: select 20 units from the lot at random and inspect them. If there is no or one defective in the sample, accept the lot. If there are three or more defectives in the sample, reject the lot. If there are exactly two defectives in the sample, draw another random sample of 25 units. Add the number defective in the first sample to the number defective found in the second sample. If the *total number defective* is two or less, accept the lot. If there are three or more defectives, reject the lot.

The motivation to use double sampling is economics; on the average double sampling plans require less total inspection than their single sampling counterparts. The reason to *not* use them is administrative; double sampling plans are a bit more confusing than single sampling plans. Also, the fact that the second sample is sometimes selected and sometimes not inspected makes the inspection workload difficult to manage.

Multiple Sampling Plan. Sampling plan where the acceptance decision may not be reached until three or more samples have been inspected. The samples are usually the same size. A multiple sampling plan with k stages works as follows:

Sample number	Sample size	Accept no.	Reject no.
1	n_1	c_1	r_1
2	n_2	c_2	r_2
.	.	.	.
.	.	.	.
.	.	.	.
k	n_k	c_k	$c_k + 1$

This sampling plan works in the same way as the double sampling plan described above except that there are more than two stages. In general, multiple sampling requires less total inspection than either single or double sampling, at the cost of additional administrative confusion.

Sequential Sampling Plan. Sampling plan where an acceptance decision is made on the inspection of each unit in the sample. Figure 8.3 illustrates the concept behind a sequential plan.

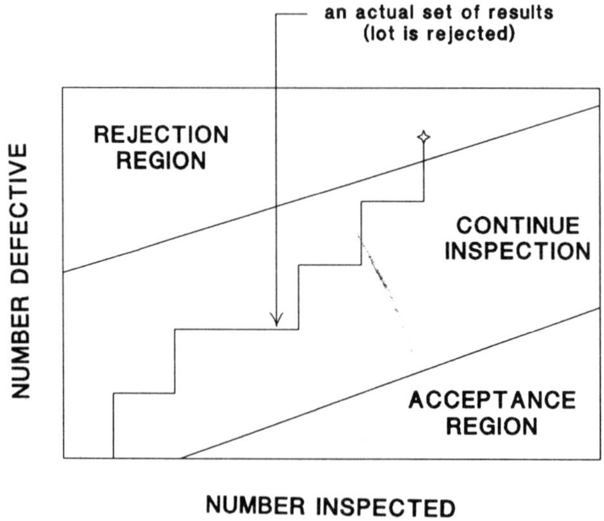

Figure 8.3 Sequential sampling plan.

Single Sampling Plans

Calculating Operating Characteristics

Operating characteristics and the OC curve are the most important tool for evaluating a particular sampling plan. Many of the items of concern to both producer and consumer can be obtained directly from the OC curve, such as AQL, LTPD, producer's risk, and consumer's risk. The remaining items can be calculated from the OC curve. We will learn a method for calculating operating characteristics that use tables of the Poisson distribution. The worksheet and instructions in Figure 8.4 make this a very simple task.

Example. You've been told by a supplier that they intend to use the following sampling plan for your product:

1. Sample 100 units from every lot.
2. Accept the lot if one or fewer defectives are found in the sample.

Your lots are typically 1000 units each. You are concerned about the performance of this sampling plan for quality levels ranging from 0.5% ($p = .005$) to 5% defective ($p = .05$). The completed worksheet is shown in Figure 8.5, and the OC and AOQ curves are shown in Figure 8.6.

Discussion

Producers are usually interested in the quality level they must achieve to have the vast majority of their lots accepted (i.e., the AQL). Typically, we look for quality levels accepted 95% of the time or more. We can see that for this plan the acceptable quality level is something better than 0.5% defective. A more precise acceptable quality level can be found by entering the Poisson table (Table 2) in the column for $c = 1$ and finding the row corresponding to a probability of acceptance of 95%. Doing this, we find $P_A = .951$ in the row for $np = 0.35$. The AQL is found by dividing the np number by the sample size (e.g., AQL = 0.35/100 = 0.0035 = 0.35%). Consumers will be more interested in how the plan protects them from poor quality. The operating characteristics curve shows that this sampling plan provides good protection from individual lots that are 4% defective. Also, if this supplier provides a steady flow of lots, the average quality after sorting of rejected lots will be no worse than 0.75% defective.

[ ] = Sample size	[ ] = Accept number
[ ] = Lot size	

fraction defective or defects per unit	np or nu	Probability of acceptance	Average Outgoing Quality
(1)	(2)	(3)	(4)

Figure 8.4 Worksheet for computing operating characteristics for single sampling plans.

INSTRUCTIONS FOR USING WORKSHEET

1. Record the sample size (n), lot size (N), and accept number (c) in the appropriate box. Since this is a single sampling plan the reject number is r = c +1.

2. Enter the lot quality levels of interest in column (1). When evaluating defectives, use fractions, not percentages.

3. Multiply the values in column (1) by the sample size, n, and record the results in column (2).

4. Table 2 in the appendix is indexed across the columns by the accept number, c, and down the rows by the numbers recorded in column (2). The numbers in the table are the probability of acceptance, P_A. Find the values for column (3) from the table.

5. Complete column (4), AOQ, by using the formula below

$$AOQ = p \times P_A \times \left(1 - \frac{\text{sample size}}{\text{lot size}}\right) \qquad (35)$$

where p is the value in column (1), P_A is the value in (3), and the sample size and lot size are given.

6. Plot the OC curve with column (1) along the bottom and column (3) on the vertical axis.

7. Plot the AOQ curve with column (1) along the bottom and column (4) on the vertical axis. Note that the AOQL can be approximately found as the largest value in column (4). The AOQL can also be calculated as (Thomas, 1955)

$$AOQL = y\ (1/n - 1/N) \qquad (36)$$

Where y is a factor from Table 7 in the appendix, N is the lot size, and n is the sample size.

Figure 8.4 (*Continued*)

100	= Sample size	1	= Accept number
1,000	= Lot size		

fraction defective or defects per unit	np or nu	Probability of acceptance	Average Outgoing Quality
(1)	(2)	(3)	(4)
.005	0.50	.910	.0041*
.010	1.00	.736	.0066
.015	1.50	.558	.0075
.020	2.00	.406	.0073
.024	2.40	.308	.0067
.030	3.00	.199	.0054
.036	3.60	.126	.0041
.040	4.00	.092	.0033
.046	4.60	.056	.0023
.050	5.00	.040	.0018

* Example: AOQ = .005 x .910 x (1- 100/1000) = .004095, round to .0041.

AOQL = .841/100 - .841/1000 = .00841-.000841 = .007569.

Note that this is very close to the .0075 we found in the table at p = .015.

Figure 8.5 Example of evaluating a single sampling plan design.

Designing Your Own Single Sampling Plans

At times you will know the goals you wish to accomplish by sampling and will want to design a sampling plan that will meet your objectives. This procedure allows you to design single sampling plans with specified producer's and consumer's risks. Recall that the producer's risk is the probability of rejecting a lot that meets the AQL, and the consumer's risk is the probability of accepting a lot that is at the RQL. The procedure used to design these plans involves the use of the Poisson table. The worksheet and instructions in Figure 8.7 are used.

Example 2: Designing a Single Sampling Plan That Meets Predefined Consumer and Producer Risks. Design a sampling plan

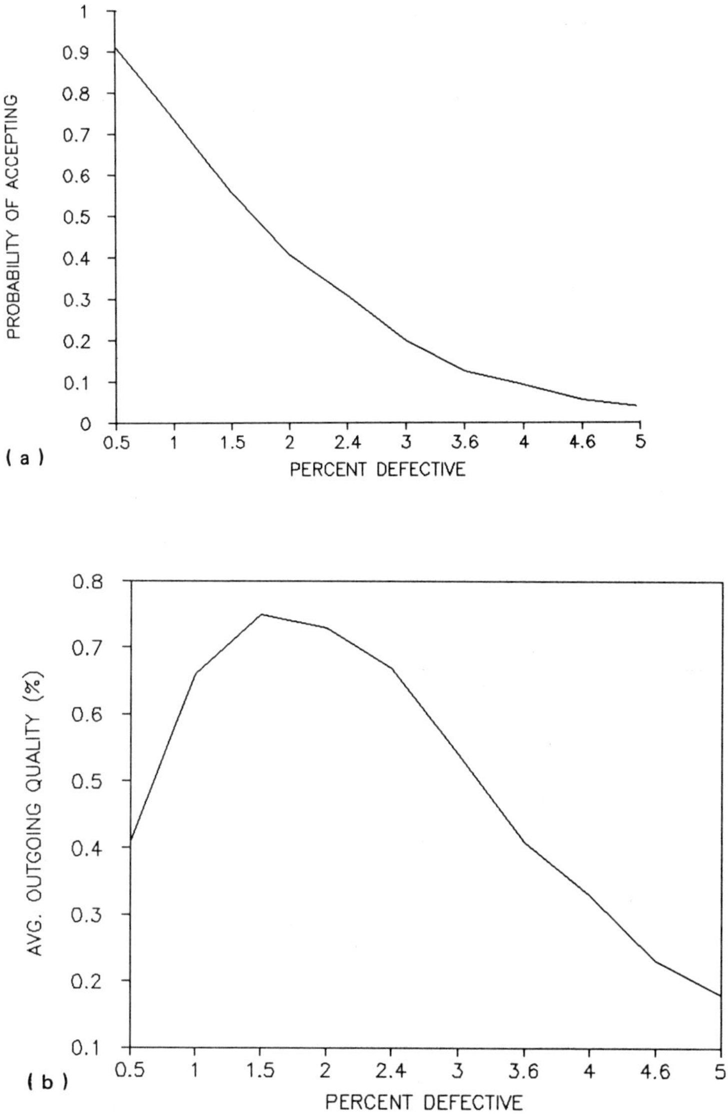

Figure 8.6 The OC and AOQ curve for the sampling plan in figure 8.5.

TARGET PARAMETERS

AQL	P_A at AQL	RQL	P_A at RQL

DISCRIMINATION = $\frac{\text{RQL}}{\text{AQL}}$ = ☐

(1) accept no.	(2) np value at AQL P_A	(3) np value at RQL P_A	(4) plan ratio (3) ÷ (2)
0			
1			
2			
3			
4			
5			
6			
7			
8			
9			
10			
11			
12			
13			
14			
15			

Figure 8.7 Single sampling plan worksheet for specified producer and consumer risks.

INSTRUCTIONS

1. Determine the target parameters, these are usually negotiated between the supplier and the consumer.
2. Calculate the desired Discrimination by dividing the Reject Quality Level (RQL) by the Acceptable Quality Level (AQL).
3. Enter table 2 of the appendix in the column accept number 0. Go down to the first row where the probability is less than or equal to the target probability of accepting AQL quality. The row label is the np value for this probability. Record the np value in column 2 of the worksheet.
4. Remain in the same column and go down to the row where probability is less than or equal to the target probability of accepting RQL quality. Record the np value for this probability in column 3 of the worksheet.
5. Divide the value in column 3 by the value in column 2 and record the result in column 4. Compare this value to the Discrimination value calculated in step 2.
   a. If the value in column 4 is larger than the Discrimination value, repeat steps 3 and 4 for the next larger accept number.
   b. If the value in column 4 is equal to or greater than the Discrimination value, proceed to step 6.
6. The accept number is the number in column 1 for the value in column 4 that is nearest to the Discrimination value computed in step 2. Compute the sample size as follows:

$$n_1 = \frac{np_{\text{at AQL}}}{\text{AQL}} \qquad (37)$$

$$n_2 = \frac{np_{\text{at RQL}}}{\text{RQL}} \qquad (38)$$

$$n = \frac{n_1 + n_2}{2} \qquad (39)$$

Round the value of n to the nearest whole number. This is the single sampling plan closest to meeting your established requirements.

Figure 8.7 *(Continued)*

that has a 95% probability of accepting lots that are 1% defective and a 5% probability of accepting lots that are 4% defective. A single sampling plan worksheet for this example is shown in Figure 8.8.

Mil-Std-105D (and ANSI/ASQC Z1.4-1980)

Introduction

Mil-Std-105D is a Department of Defense standard for inspection by attributes. It includes sampling plans and procedures. The ANSI/ASQC equivalent standard is Z1.4.

TARGET PARAMETERS

AQL	P_A at AQL	RQL	P_A at RQL
.01	.95	.04	.05

DISCRIMINATION = $\frac{\text{RQL}}{\text{AQL}}$ = 4

(1) accept no.	(2) np value at AQL P_A	(3) np value at RQL P_A	(4) plan ratio (3) ÷ (2)	
0	.0513	2.996	58.4	
1	.355	4.744	13.4	
2	.818	6.296	7.7	
3	1.366	7.754	5.7	
4	1.970	9.154	4.6	
5	2.613	10.513	4.0	*

$n_1 = 2.613 \div .01 = 261.3$

$n_2 = 10.513 \div .04 = 262.825$

$n = (261.3 + 262.825) \div 2 = 524.125 \div 2 = 262$

Thus, the single sampling plan closest to meeting our requirements is

n = 262
c = 5

Figure 8.8 Example of using a single sampling plan worksheet for specified producer and consumer risks.

Sampling plans in Mil-Std-105D are intended for use with an ongoing series of lots or batches. Mil-Std-105D can be applied to either defectives or defects. Excerpts from Mil-Std-105D are given in the Appendix.

How to Use Mil-Std-105D

Determine the AQL. The AQL must be one of the "preferred AQL's" in the master tables.

Determine the inspection level. Use general inspection level II unless there is a good reason to select some other level.

Determine the type of inspection to be used: single, double, or multiple.

Determine the switching level: reduced, normal, or tightened. You always begin at normal and follow the switching procedures described in Mil-Std-105D.

Determine the size of the inspection lot.

Enter Table I and find the sample size code letter for your lot size and level of inspection.

Depending on the severity and type of inspection, enter one of the master tables and find the sampling plan. A guide to the master tables is shown in Figure 8.9.

Note that there are cases where you will encounter an arrow in a master table. When this occurs, follow the arrow to the first row in the table that contains an accept (Ac) or a reject (Re) number and use the sampling plan in that row. Be sure to read the notes at the bottom of the master table for special instructions.

TYPE OF INSPECTION	SEVERITY OF INSPECTION		
	REDUCED	NORMAL	TIGHTENED
SINGLE	II-C	II-A	II-B
DOUBLE	III-C	III-A	III-B
MULTIPLE	IV-C	IV-A	IV-B

Figure 8.9 Master table selection guide for Mil-Std-105D

Draw the samples at random from the lot and inspect each item in the sample. Determine the acceptability of the lot by comparing the results to the acceptance/rejection criteria given in the master table.

Disposition of the lot. This may mean acceptance, rework, scrap, etc., depending on your sample results and acceptance criteria. Be sure that you have a record of your disposition. The record will be used for switching as well as for quality control purposes.

Where appropriate, take action on the process to effect a quality improvement.

Dodge-Romig Sampling Tables

Dodge-Romig tables are attribute acceptance sampling tables designed for use with ongoing processes that have known process averages. There are tables for either single sampling or double sampling. Sampling plans in the Dodge-Romig tables are designed to minimize the average total inspected. There are two classes of sampling plans: AOQL constrained and LTPD constrained.

To use the Dodge-Romig tables, the following information must be provided:

The AOQL, for AOQL-constrained plans, or the LTPD, for LTPD-constrained plans. For the LTPD-constrained plans, you have a choice of probabilities of acceptance at the LTPD quality level: 5% or 10%.

The process average fraction defective.

The lot size.

The type of sampling to be used (single or double sampling).

General Comments on Acceptance Sampling

It can be proved mathematically that acceptance sampling should never be used for processes in statistical control (Mood, 1943). There are two reasons for this: (1) if the process is in statistical control there is no correlation between the number of defectives in the sample and the number of defectives remaining in the lot, and (2) the most economical amount of inspection for a process in statistical control is either 100% or 0%, never a partial sample of the lot (Deming, 1982, Chapter 13). Acceptance sampling is a detection approach to quality control, while SPC is an approach that prevents problems. Thus, concentrate on getting statistical control first—then you will need acceptance sampling only for isolated lots or other special situations such as purchasing from a central distribution warehouse where no process history is available.

There are many occasions when rejected lots are reinspected. This is poor practice, and it creates a sampling plan that has very high consumer risks.

It is often difficult to draw random samples, but it is also very important. If you don't plan to draw your samples at random, don't dignify the inspection process by pretending that a sampling plan will somehow make it "scientific."

If you are sampling for critical defects, the accept number should always be zero. Also, sampling should be used only to verify that a prior 100% inspection or test was effective. The reasons for this are both ethical and legal.

If, while sampling, you observe defects that are not part of your sample, you must identify the defective units and remove them from the inspection lot. You cannot knowingly ship a defective product, regardless of whether it is part of your sample. Of course, the acceptability of the lot is determined only by the units in the sample. Keep in mind that you cannot knowingly exclude a unit from your sample because it is defective (i.e., don't remove obvious defectives before drawing the sample).

SUMMARY

This chapter has introduced the basic terminology of acceptance sampling. The underlying concepts were presented, especially those related to sampling risk. Different types of sampling plans (single, double, multiple, and sequential) were explained. The design and evaluation of single sampling plans were discussed and examples were given. Military standard Mil-Std-105D was introduced and briefly discussed. Finally, the shortcomings of acceptance sampling as a quality control method were outlined along with reasons why SPC is preferred.

Not discussed was the mathematics of acceptance sampling. No proofs or derivations were given for any of the methods introduced. Several types of acceptance sampling were omitted, such as continuous sampling, acceptance sampling by variables, skip-lot plans, chain sampling, and Bayesian sampling. The design of double, multiple, and sequential sampling plans was not discussed. Finally, a variety of standards and specifications on acceptance sampling published by the

Department of Defense and private organizations exist but were not discussed.

RECOMMENDED READING LIST

49, 53, 55–60, 62–64.

9

Designed Experiments and Taguchi Methods

Designed experiments play an important role in quality improvement. This chapter introduces the basic concepts involved and contrasts the statistically designed experiment with the "one variable at a time" approach that has been used traditionally. Also briefly discussed are the concepts involved in Taguchi methods, statistical methods named after their chief proponent, Dr. Genich Taguchi. Taguchi methods involve application of designed experiments to quality in the design, production, and use phases of a product's life cycle. Taguchi also introduced a number of new concepts in quality engineering, such as new applications of loss functions to quality. Readers interested in a more in-depth treatment of the subjects discussed in this chapter are referred to Taguchi (1987) and Montgomery (1984).

TERMINOLOGY

Designed experiment. An experiment where one or more variables (called independent variables) that are believed to have an effect on the experimental outcome are identified and manipulated according to a predetermined plan. Data collected from a designed experiment can be analyzed statistically to determine the effect of an independent variable or a combination of independent variables. An experimental plan must also include provisions for dealing with extraneous variables, that is, variables not explicitly identified as independent variables.

Response variable. The variable being investigated, also called the dependent variable.

Primary variables. The controllable variables believed most likely to have an effect. These may be quantitative, such as temperature, pressure, or speed, or qualitative, such as vendor, production method, or operator.

Background variables. Variables, identified by the designers of the experiment, which may have an effect but either cannot or should not be deliberately manipulated or held constant. The effect of background variables can contaminate primary-variable effects unless they are properly handled. The most common method of handling background variables is blocking (described later in this chapter).

"Common causes" or experimental error. In any given experimental situation, a great many variables may be potential sources of variation—so many, in fact, that no experiment could be designed that deals explicitly with every possible source of variation. The variables that are not dealt with explicitly are analogous to the common causes of variation described in Chapter 7. They represent the "noise level" of the process, and their effects are kept from contaminating the primary-variable effects by randomization. Randomization is a procedure that assigns test units to test conditions in such a way that any given unit has the same probability as any other unit of being processed under a given set of test conditions.

Fixed-effects model. An experimental model in which all factor levels of interest are studied. For example, if there are three different materials, all three are included in the experiment.

Random-effects model. Experimental model in which the levels of factors evaluated by the experiment represent a sample of all possible levels—for example, if we have three different materials but only use two materials in the experiment.

Mixed model. Experimental model with both fixed and random effects.

Completely randomized design. Experimental plan where the order in which the experiment is performed is completely random. For example:

Level	Test sequence number
A	7,1,5
B	2,3,6
C	8,4

Randomized block design. Experimental design in which the experimental observations are divided into "blocks" according to some criterion. The blocks are filled sequentially, but the order within each block is filled randomly. For example, assume we are conducting a painting test with different materials, material A and material B. We have four test pieces of each material. Ideally, we would like to clean all of the peices at the same time to assure that the cleaning process doesn't have an effect on our results. But what if our test requires that we use a cleaning tank that cleans two test pieces at a time? The tank load then becomes a "blocking factor." We will have four blocks, which might look like this:

Material	Tank load	Test piece number
A	1	7
B		1
B	2	5
A		2
B	3	3
A		6
B	4	4
A		8

Since each material appears exactly once per cleaning tank load, the material totals or averages can be compared directly. The reader should be aware that statistical designs exist to handle more complicated "unbalanced designs" (for details see Montgomery, 1984).

Replication. Collection of more than one observation for the same set of experimental conditions.

Interaction. Condition in which the effect of one factor depends on the level of another factor. Interaction is illustrated in Figure 9.1. The chart on the left illustrates a situation where the effect of varying factor A depends on the level of factor B. If B is at its low level, then increasing

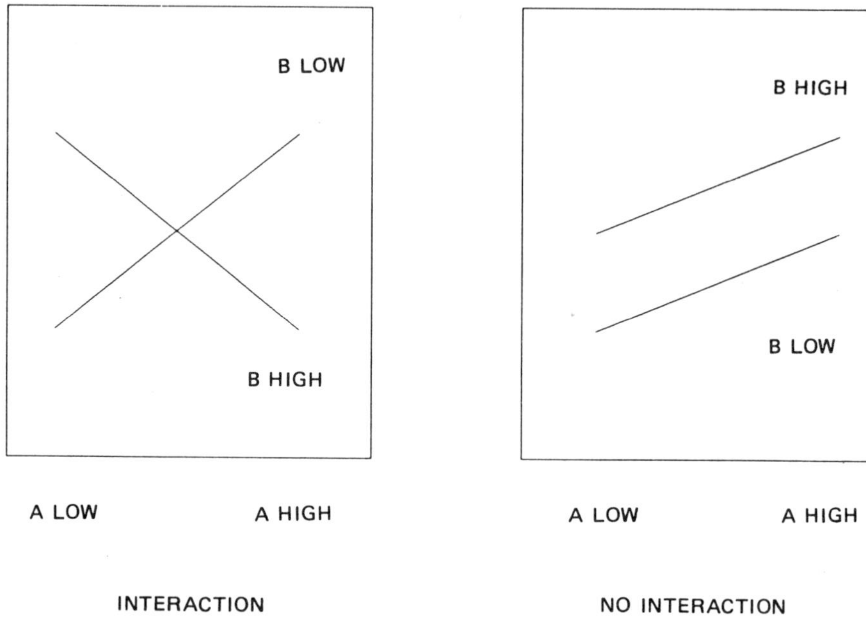

Figure 9.1 Illustration of interaction.

factor A from its low level to its high level results in increasing the response variable. However, increasing factor A when factor B is at its high level has the opposite effect. The chart on the right illustrates no interaction, in that varying factor A has the same effect regardless of the level of factor B.

THE "TRADITIONAL APPROACH" VERSUS STATISTICALLY DESIGNED EXPERIMENTS

The traditional approach is to hold all variables except one constant. The statistically designed experiment usually involves varying two or more variables simultaneously. The advantages of the statistical approach are twofold:

1. Interactions, such as those illustrated above, can be detected and measured. Failure to detect interactions is a major flaw in the "one variable at a time" approach.

2. Each value does the work of several values. A properly designed experiment allows you to use the same observation to estimate several different effects. This translates directly to cost savings when using the statistical approach.

TAGUCHI METHODS

This section presents some of the special concepts introduced by Dr. Genichi Taguchi of Japan. A complete discussion of Taguchi's approach to designed experiments is beyond the scope of this book. However, many of Taguchi's ideas are useful in that they present an alternative way of looking at quality in general.

Quality is defined as the loss imparted to the society from the time a product is shipped (Taguchi, 1986). Taguchi divides quality control efforts into two categories: on-line quality control and off-line quality control.

On-line quality control involves diagnosis and adjustment of the process, forecasting and correction of problems, inspection and disposition of product, and follow-up on defectives shipped to the customer.

Off-line quality control methods are quality and cost control activities conducted at the product and process design stages in the product development cycle. Kackar (1985) lists three major aspects of off-line quality control:

System design: the process of applying scientific and engineering knowledge to produce a basic functional prototype design. The prototype model defines the initial settings of product or process design characteristics.

Parameter design: an investigation conducted to identify settings that minimize (or at least reduce) the performance variation. A product or a process can perform its intended function at many settings of its design characteristics. However, variation in the performance characteristics may change with different settings. This variation increases both product manufacturing and lifetime costs. The term parameter design comes from an engineering tradition of referring to product characteristics as product parameters. An exercise to identify optimal parameter settings is therefore called parameter design.

Tolerance design: a method for determining tolerances that minimize the sum of product manufacturing and lifetime costs.

> The final step in specifying product and process designs is to determine tolerances around the nominal settings identified by parameter design. It is still a common practice in industry to assign tolerances by convention rather than scientifically. Tolerances that are too narrow increase manufacturing costs, and tolerances that are too wide increase performance variation and the lifetime cost of the product.

Expected loss refers to the monetary losses an arbitrary user of the product is likely to suffer at an arbitrary time during the product's life span because of performance variation. Taguchi advocates modeling the loss function so that the issue of parameter design can be made more concrete. The most often used model of loss is the quadratic loss function illustrated in Figure 9.2.

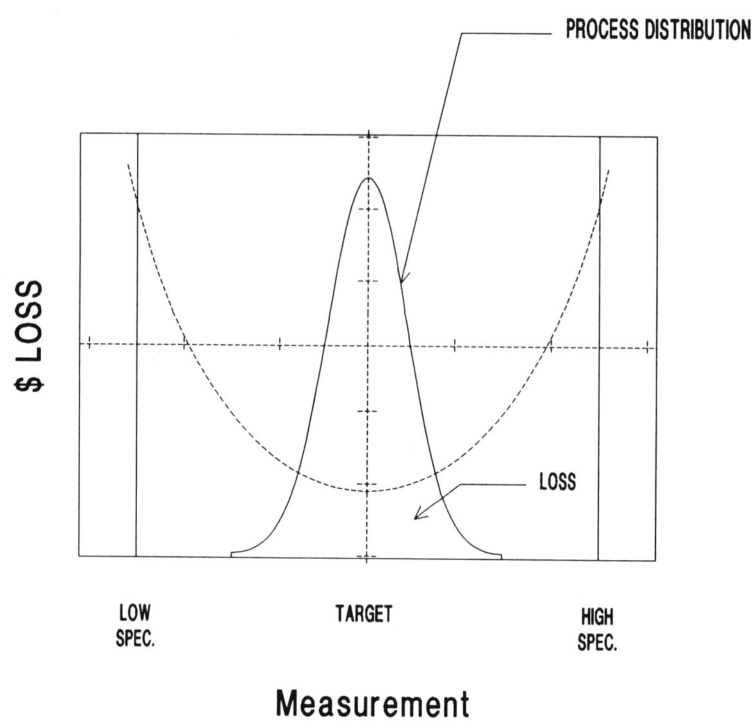

Figure 9.2 Quadratic loss function.

The traditional American model of loss is illustrated in Figure 9.3. Note the contrast between the quadratic loss function and the conceptual loss function traditional in America.

The interpretation of Figure 9.3 is that there is no loss as long as a product meet requirements. A part that is just barely inside specifications is as good as one that is exactly nominal. A part that is beyond the specification limit incurs a loss. Parts that deviate a great deal from specifications incur the same loss as those just barely outside (usually scrap or rework).

Noise is the term used to describe all the variables, except design parameters, that cause performance variation during a product's life span and across different units of the product. Sources of noise are classified as either external sources or internal sources.

External sources of noise are variables external to a product that affect the product's performance.

Internal sources of noise are the deviations of the actual characteristics of a manufactured product from the corresponding nominal settings.

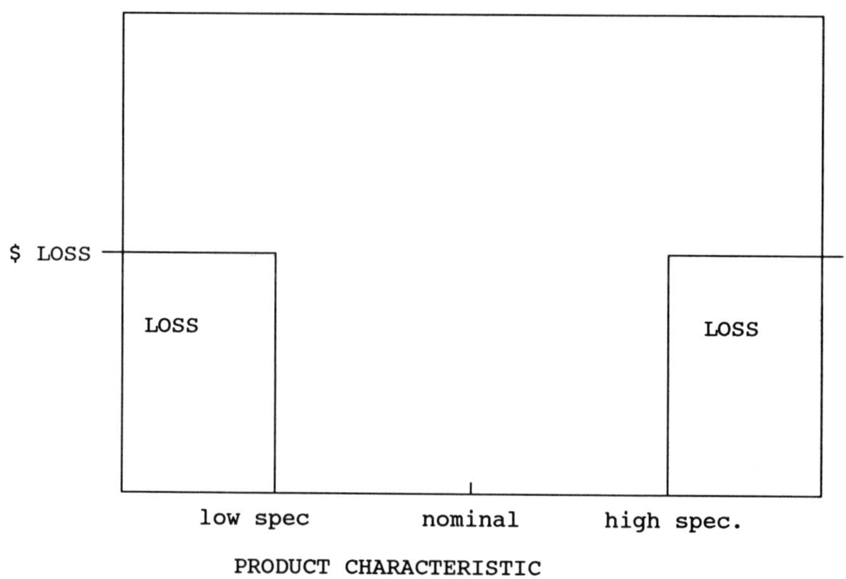

Figure 9.3 Traditional American view of loss.

Performance statistics estimate the effect of noise factors on the performance characteristics. Performance statistics are chosen so that maximizing the performance measure will minimize expected loss. Many performance statistics used by Taguchi use "signal to noise ratios" which account jointly for the levels of the parameters and the variation of the parameters.

Summary of the Taguchi Method

The Taguchi method for identifying settings of design parameters that maximize a performance statistic is summarized by Kackar (1985) as follows:

Identify initial and competing settings of the design parameters, and identify important noise factors and their ranges.
Construct the design and noise matrices, and plan the parameter design experiment.
Conduct the parameter design experiment and evaluate the performance statistic for each test run of the design matrix.
Use the values of the performance statistic to predict new settings of the design parameters.
Confirm that the new settings indeed improve the performance statistic.

SUMMARY

The advantages of statistical design over the classical one-at-a-time approach were presented. The basic concepts and definitions of experimental design were introduced. The fundamental concepts of Taguchi's method of quality control were discussed.

The chapter did not cover the mathematics of experimental design. No formulas were provided, nor was any detailed framework provided for conducting a designed experiment or analyzing the data from one.

RECOMMENDED READING LIST

65–70.

10

Reliability

While common usage of the term reliability varies, its technical meaning is quite clear. Reliability is defined as the probability that a product or system will perform a specified function for a specified time without failure. For the reliability figure to be meaningful, the operating conditions must be very carefully and completely defined. Although reliability analysis can be applied to just about any product or system, in practice it is normally applied only to complex products. Formal reliability analysis is routinely used for commercial products, such as automobiles, as well as military products like missiles.

TERMS

MTBF. Mean time between failures. When applied to repairable products, this is the average time a system will operate until the next failure.

MTTF or MTFF. Mean time to first failure. This is the measure applied to systems that can't be repaired during their mission. For example, the MTTF would be relevant to the Voyager spacecraft.

Failure rate. The number of failures per unit of stress. The stress can be time (e.g., machine failures per shift), load cycles (e.g., wing fractures per 100,000 deflections of 6 inches), impacts (e.g., ceramic cracks per 1000 shocks of 5 g's each), or one of a variety of other stresses. The failure rate is computed by dividing 1 by the MTTF or MTBF.
MTTR. Mean time to repair.
Availability. The proportion of time a system is operable. This is relevant only for systems that can be repaired. Availability is given by the equation

$$A = \frac{\text{MTBF}}{\text{MTBF} + \text{MTTR}} \tag{40}$$

b_{10} life*. The life value at which 10% of the population has failed.
b_{50} life. The life value at which 50% of the population has failed. Also called the median life.
Fault tree. Diagram used to trace symptoms to their root causes. *Fault tree analysis* is the process involved in constructing a fault tree.
Derating. Assigning a product to an application that is at a stress level less than the rated stress level for the product. This is analogous to providing a safety factor.
Censored test. A life test in which some units are removed before the end of the test period, even though they have not failed.
Maintainability. A measure of the ability to place a system that has failed back in service. Figures of merit include availability and mean time to repair.
Bathtub curve. The life cycle of many complex systems can be modeled using a curve called the bathtub curve, which is shown in Figure 10.1.

The bathtub curve accurately describes failure rates that appear in the real world with complex systems. The first portion of the bathtub curve, called the infant mortality period, is characterized by a high but rapidly declining rate of failure.† The infant mortality period is not a good one for routine operation of a system, and typically these early

---

*The terms b_{10} life and b_{50} life are commonly applied to the reliability of bearings.
†As you might suspect, the term infant mortality is a reference to the death rate of human infants. In fact, human beings *are* complex systems, and our life cycle is accurately modeled with the bathtub curve.

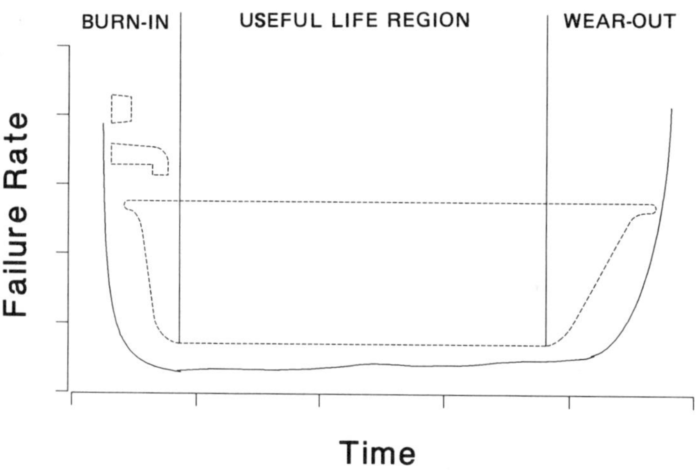

Figure 10.1 Bathtub curve.

failures are weeded out by *burning in* the system. Burn-in is nothing more than running the system under conditions that simulate a high-stress operating environment for a period of time sufficient to allow the failure rate to stabilize.

The area of the bathtub curve where the failure rate is relatively constant is the useful life portion for the system. This is followed by a zone of increasing failure rates, which is known as the wear-out region for the system. Nearly all reliability analysis is conducted in the useful life, or constant failure rate, region.

MATHEMATICAL MODELS

The mathematics of reliability analysis is a subject unto itself. Most systems of practical size require the use of high-speed computers for reliability evaluation. However, an introduction to the simpler reliability models is extremely helpful in understanding the concepts involved in reliability analysis.

One statistical distribution that is very useful in reliability analysis is the exponential distribution, which is given by Equation (41):

$$R(t) = e^{-t/\mu} \qquad t \geqslant 0 \tag{41}$$

In this equation R is the system reliability, given as a probability, t is the time the system is required to operated without failure, and μ is the mean time to failure or mean time between failures for the system. The exponential distribution applies to systems operating in the constant-failure-rate region of the bathtub curve, which is where most systems are designed to operate.

RELIABILITY APPORTIONMENT

Since reliability analysis is commonly applied to complex systems, it is logical that most of these systems are composed of smaller subsystems. Apportionment is the process involved in allocating reliability objectives among separate elements of a system. The final system must meet the overall reliability goal. Apportionment is something of a hybrid of project management and engineering disciplines.

The process of apportionment can be simplified if we assume that the exponential distribution is a valid model. The exponential distribution has a property that allows the system failure rate to be computed as the reciprocal of the sum of the failure rates of the individual subsystems. Let's look at an example. Table 10.1 shows the apportionment for a home entertainment center. The complete system is composed of a tape deck, television, compact disk unit, and phonograph. Assume that the overall objective is a reliability of 95% at 500 hours of operation.

The apportionment could continue even further; for example, we could apportion the drive reliability to pulley, engagement head, belt, capstan, etc. The process ends when it has reached a practical limit.

Table 10.1 Reliability Apportionment for a Home Entertainment System

Subsystem	Reliability	Unreliability	Failure rate	Objective
Tape deck	.990	.010	.00002	49,750
Television	.990	.010	.00002	49,750
Compact disk	.985	.015	.00003	33,083
Phonograph	.985	.015	.00003	33,083
	.950	.050		
		Tape deck subsystem		
Drive	.993	.007	.000014	71,178
Electronics	.997	.003	.000006	166,417

The column labeled "Objective" in Table 10.1 gives the minimum acceptable mean time between failures for each subsystem. Values of the MTBF below this will cause the entire system to fail its reliability objective. Note that the required MTBFs are huge compared to the overall objective of 500 hours for the system as a whole. This happens partly because of the fact that the reliability of the system as a whole is the *product* of the subsystem reliabilities, which requires the subsystems to have much higher reliabilities than the complete system.

Reliability apportionment is very helpful in identifying design weaknesses. It is also an eye-opener for management, vendors, customers, and others to see how the design of an entire system can be dramatically affected by one or two unreliable elements.

RELIABILITY PREDICTION

Series Reliability

A system is in series if all of the individual elements must function for the system to function. A series system is shown in Figure 10.2.

In Figure 10.2 the system is composed of two elements, A and B. Both A and B must function correctly for the system to function correctly. The reliability of this system is equal to the product of the reliabilities of A and B, in other words:

$$R_s = R_A \times R_B \tag{42}$$

For example, if the reliability of A is .99 and the reliability of B is .92, then $R_s = .99 \times .92 = .9108$. Note that with this configuration R_s is always less than the *minimum* of R_A or R_B. This implies that the best way to improve the reliability of the system is to work on the system component that has the lowest reliability.

Parallel Reliability

A system is said to have a parallel configuration if only one component needs to function for the system to function. A parallel system is shown in Figure 10.3.

With parallel configurations there is *redundancy*; that is more than one component or subsystem can perform the same function. The system shown in Figure 10.3 will function as long as A, B, or C has not

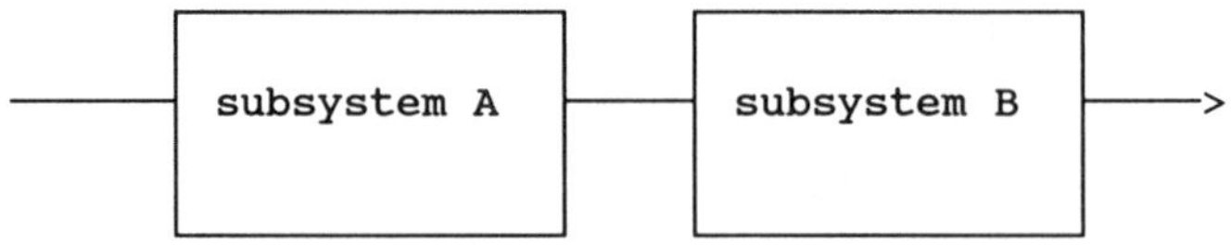

Figure 10.2 Series Configuration.

failed. The reliability of this type of configuration is computed with the equation

$$R_s = 1 - (1 - R_A)(1 - R_B)(1 - R_C) \quad (43)$$

For example, if $R_A = .90$, $R_B = .95$ and $R_C = .93$ then $R_s = 1 - (.1 \times .05 \times .07) = 1 - .00035 = .99965$.

With parallel configurations the system reliability is always better than the best subsystem reliability. Thus, when trying the improve the reliability of a parallel system you should first try to improve the reliability of the *best* component. This is exactly the opposite of the approach taken to improve the reliability of series configurations.

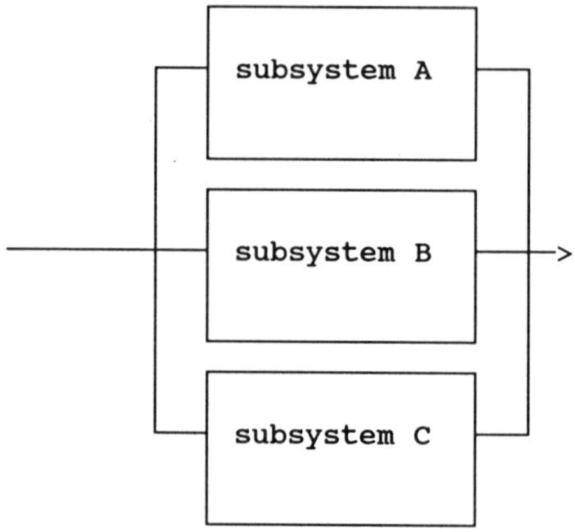

Figure 10.3 Parallel configuration.

Series-Parallel Systems

Systems often combine both series and parallel elements. Figure 10.4 illustrates this approach. The reliability of this system is evaluated in two steps:

1. Convert B-C into a single element by solving the parallel configuration.
2. Solve A, B-C, D as a series configuration.

Assume $R_A = .990, R_B = .800, R_C = .700$, and $R_D = .995$. Then the system reliability R_s is found as follows:

1. $R_{B,C} = 1 - .2 \times .3 = .940$
2. $R_s = .990 \times .940 \times .995 = .926$.

Standby Systems

Standby systems are systems in which one or more elements are inactive until some other element or elements in the system fail. In addition to the standby elements, there are sensors to detect the conditions required to place the inactive element in service and some switching mechanism for engaging the inactive element. Figure 10.5 depicts a standby system.

The reliability of a standby system is computed with the formula

$$R_s = R_A + (1 - R_A)R_{SE}R_{SW}R_B \tag{44}$$

In this equation R_{SE} is the reliability of the sensor and R_{SW} is the reliability of the switch.

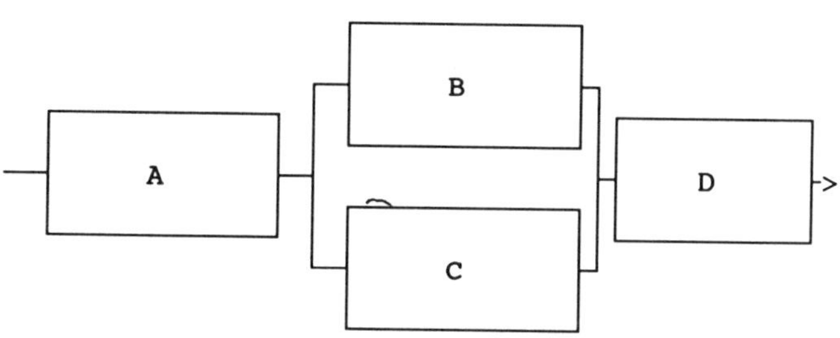

Figure 10.4 Series-parallel system.

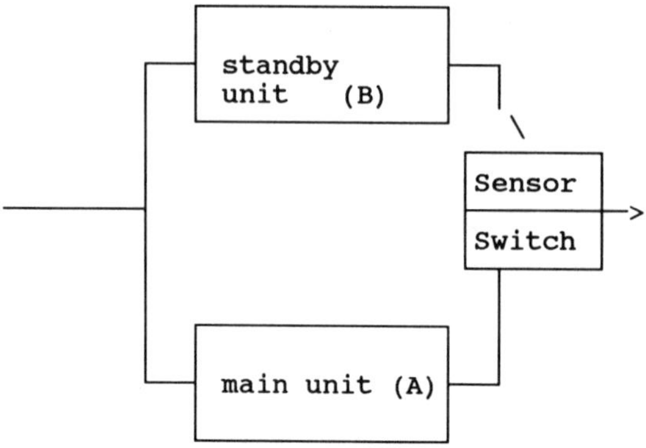

Figure 10.5 Standby system.

Design Review

Design reviews are conducted by specialists, usually working on teams. Designs are, of course, reviewed on an ongoing basis as part of the routine work of a great number of people. However, the term as used here refers to the formal design review process. The purpose of design review is threefold:

1. Determine if the product will actually work and meet the customer's requirements.
2. Determine if the new design is producible and inspectable.
3. Determine if the new design is maintainable.

Design review should be conducted at various times in the design and production process. Review should take place on preliminary design sketches, after prototypes have been designed, after prototypes have been built and tested, as developmental designs are released, etc. Designs subject to review should include parts, subassemblies, and assemblies.

Failure Mode and Effect Analysis (FMEA)

Failure mode and effect analysis, or FMEA, is an attempt to delineate all possible failures and their effect on the system. The objective is to classify failures according to their effect. FMEA provides an excellent basis for classification of characteristics.

When engaged in FMEA, it is wise to bear in mind that the severity of failure is not the only important factor; one must also consider the *probability* of failure. As with Pareto analysis, one purpose of FMEA is to direct the available resources toward the most promising opportunity. An extremely unlikely failure, even a failure with serious consequences, may not be the best place to concentrate reliability improvement efforts. Decision analysis methods may be helpful in dealing with this type of question.

Seven Steps in Predicting Design Reliability

1. Define the product and its functional operation. Use functional block diagrams to describe the systems. Define failure and success in unambiguous terms.
2. Use *reliability block diagrams* to describe the relationships of the various system elements (i.e., series, parallel, etc.).
3. Develop a reliability model of the system.
4. Collect part and subsystem reliability data. Some of the information may be available from existing data sources. Special tests may be required for other information.
5. Adjust data to fit the special operating conditions of your system. Use care to ensure that your "adjustments" have a scientific basis and are not simply reflections of personal opinions.
6. Predict reliability using a mathematical model. Computer simulation may also be required.
7. Verify your prediction with field data. Modify your models and predictions accordingly.

SYSTEM EFFECTIVENESS

The effectiveness of a system is a broader measure of performance than simple reliability. Three elements are involved in system effectiveness:

1. Availability
2. Reliability
3. Design capability (i.e., assuming the design functions, does it also achieve the desired result?)

System effectiveness can be measured with the equation below.

$$P_{SE} = P_A \times P_R \times P_C \tag{45}$$

In this equation, P_{SE} is the probability that the system will be effective, P_A is the availability as computed with Equation (40), P_R is the system reliability, and P_C is the probability that the design will achieve its objective.

SAFETY FACTORS

The modern evaluation of reliability takes into account the probabilistic nature of failures. With the traditional approach a safety factor would be defined using the equation

$$SF = \frac{\text{average strength}}{\text{worst expected stress}} \tag{46}$$

The problem with this approach is quite simple: it doesn't account for variation in stress or strength. The fact is that both strength and stress will vary over time, and unless this variation is dealt with explicity we have no idea what the "safety factor" really is.

The modern view is that a safety factor is the difference between an improbably high stress (the maximum expected stress, or "reliability boundary") and an improbably low strength (the minimum expected strength). Figure 10.6 illustrates the modern view of safety factors. The figure shows two *distributions*, one for stress and one for strength.

Since any strength or stress is theoretically possible, the concept of a safety factor becomes vague at best and misleading at worst. To deal intelligently with this situation we must consider *probabilities* instead of

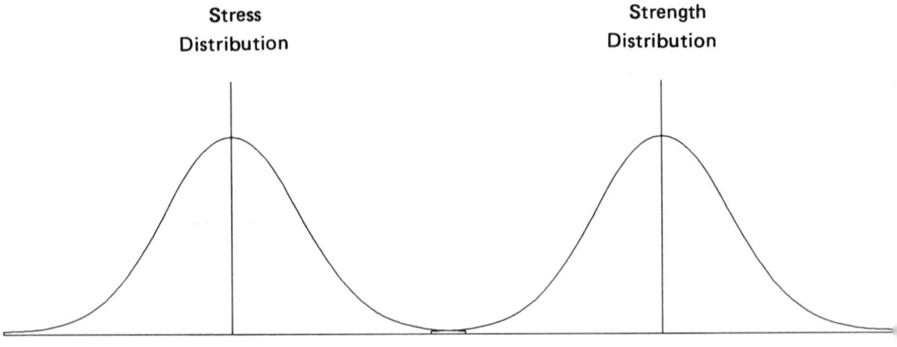

Figure 10.6 Modern view of safety factors.

possibilities. This is done by computing the probability that a stress/strength combination will occur such that the stress applied exceeds the strength. It is possible to do this since, if we have distributions of stress and strength, the difference between the two distributions is also a distribution. The average and standard distribution of the difference can be determined using statistical theory. The necessary equations are shown below.

$$\sigma_{sf}^2 = \sigma_{\text{strength}}^2 + \sigma_{\text{stress}}^2 \tag{47}$$

$$\mu_{sf} = \mu_{\text{strength}} - \mu_{\text{stress}} \tag{48}$$

In the equations above, the subscript *sf* refers to the safety factor, σ^2 refers to the variance, and μ refers to the mean.

Assume that we have normally distributed strength and stresses. Then the distribution of strength minus stress is also normally distributed with the mean and variance computed from Equations (47) and (48). Futhermore, the probability of a failure is the same as the probability that the difference of strength minus stress is less than zero.

Example. The strength of a steel rod is normally distributed with μ = 50,000# and σ = 5,000#. The steel rod is to be used as an undertruss on a conveyor system. The stress observed in the actual application was measured by strain gages and was found to have a normal distribution with μ = 30,000# and σ = 3,000#. What is the expected reliability of this system?

Solution. The mean and standard deviation of the difference is first computed using Equations (47) and (48), giving

$$\sigma_{\text{difference}}^2 = 5000^2 + 3000^2 = 34{,}000{,}000$$

$$\sigma_{\text{difference}} = \sqrt{34{,}000{,}000} = 5831$$

$$\mu_{\text{difference}} = 50{,}000^{\#} - 30{,}000^{\#} = 20{,}000^{\#}$$

We now compute Z using Equation (11) from Chapter 6 and get

$$Z = \frac{0^{\#} - 20{,}000^{\#}}{5831} = -3.43$$

Using a normal table, we now look up the probability associated with this Z value and find it is .0003. This is the probability of failure, about 1 chance in 3333. The reliability is found by subtracting this probability from 1, giving .9997. Thus, the reliability of this system is 99.97%.

SUMMARY

The basic terminology of reliability was defined. The "bathtub curve" model of reliability was introduced. Reliability apportionment was defined and an example given. Reliability prediction was briefly discussed. Failure mode and effect analysis (FMEA) was defined. The importance of design review was explained, as was the concept of system effectiveness. A modern, probabilistic view of safety factors was presented.

The chapter did not discuss the derivation of any of the models presented. The important topic of maintainability was not discussed in depth. Only simple models were presented, even though most real-world systems are extremely complex. Methods of analyzing the reliability of complex models, such as computer simulation, were not presented. Finally, the chapter did not mention the many standards and specifications for reliability.

RECOMMENDED READING LIST

24, 71–81.

11

Measurement Error

Most quality control activities involve the acquisition, analysis, interpretation, or storage of data. The success of the action taken depends on the reliability of the raw data that form the basis for informed decision making. Computer scientists coined the phrase "garbage in, garbage out" (GIGO) to describe the result of performing sophisticated analysis of erroneous input data. This chapter describes statistical methods of evaluating errors of measurement. The primary objective is to provide assurance that measurement errors are small enough that their impact on decisions becomes negligible.

EVALUATION OF MEASUREMENT ERROR

Definition. Measurement error is the difference between an actual condition and the estimate of that condition based on one or more measurements. In the case of attributes data, measurement error refers to the correctness of classifications (e.g., the percentage of acceptable parts that are rejected or the percentage of reject parts that are accepted).

Usage. Measurement error should be quantified and evaluated with respect to process performance. Effective statistical process control and quality control require data that provide a valid basis for making process control decisions. This applies to attributes or continuous variables. Thus, measurement error evaluation should be performed early in any SPC project. Control of measurement error should be an integral part of any quality control plan.

Measurement Error Concepts

Accuracy. A measurement system is said to be accurate if, on the average, the measured value is the same as the true value. The difference between the average measured value and the true value is referred to as *bias*. Bias refers to a systematic error, as contrasted with random error. In practice, "true values" are approximated by accepted reference standards. When applied to attribute inspection, accuracy refers to the ability of the attribute inspection system to produce agreement on inspection standards. Accuracy is usually controlled by *calibration*. Figure 11.1 illustrates the accuracy concept.

Repeatability or Precision. A measurement system is said to be *repeatable* if it always produces the same result. Variation obtained when the measurement system is applied repeatedly under the same conditions is usually caused by conditions inherent in the measure-

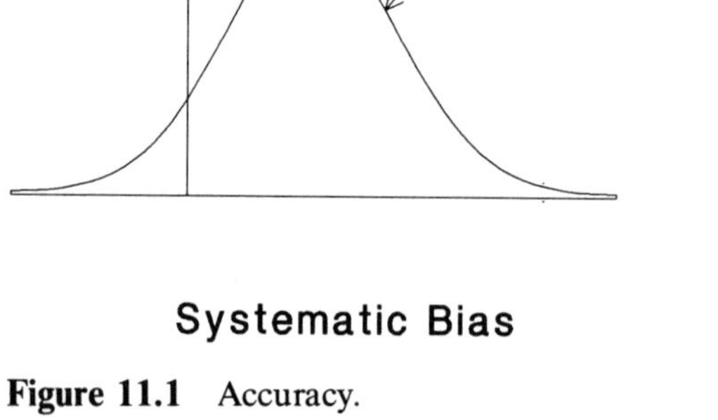

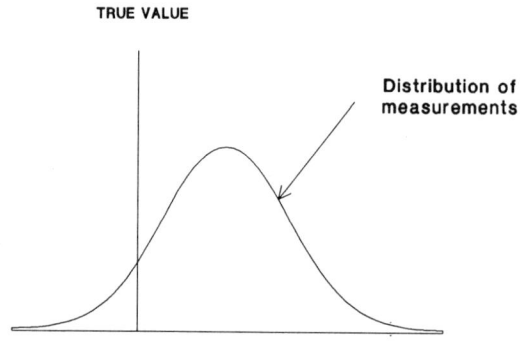

Figure 11.1 Accuracy.

ment system. This is sometimes referred to as random variation. Repeatability error is illustrated in Figure 11.2.

Reproducibility. If different individuals using the same measurement system get the same results, then the results are said to be reproducible (Figure 11.3). Note that this definition is different from the classical scientific definition of reproducibility.

Stability. A system is said to be stable if the results are the same at different points in time (Figure 11.4).

Evaluating Measurement Accuracy

Gage accuracy can be analyzed by taking several repeat readings on items whose true value (size, weight, etc.) is known. In the typical company, gage accuracy is carefully controlled by a *calibration system*. Also, most inspection operations have a *standard* that is used periodically to reverify the accuracy of the gage. When attribute measures are used, the "calibration" activity is *training*. All personnel who will be asked to pass

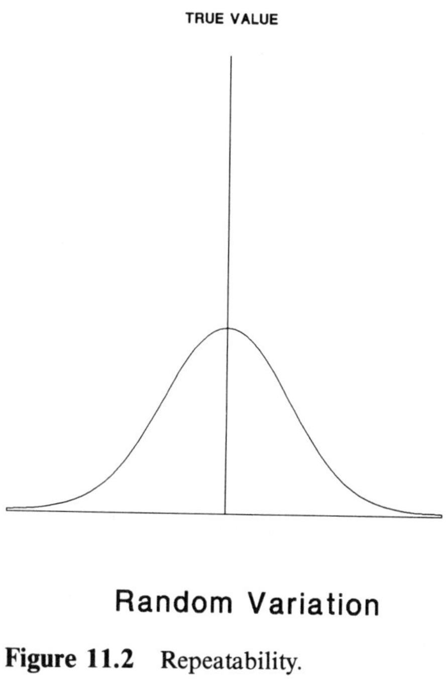

Figure 11.2 Repeatability.

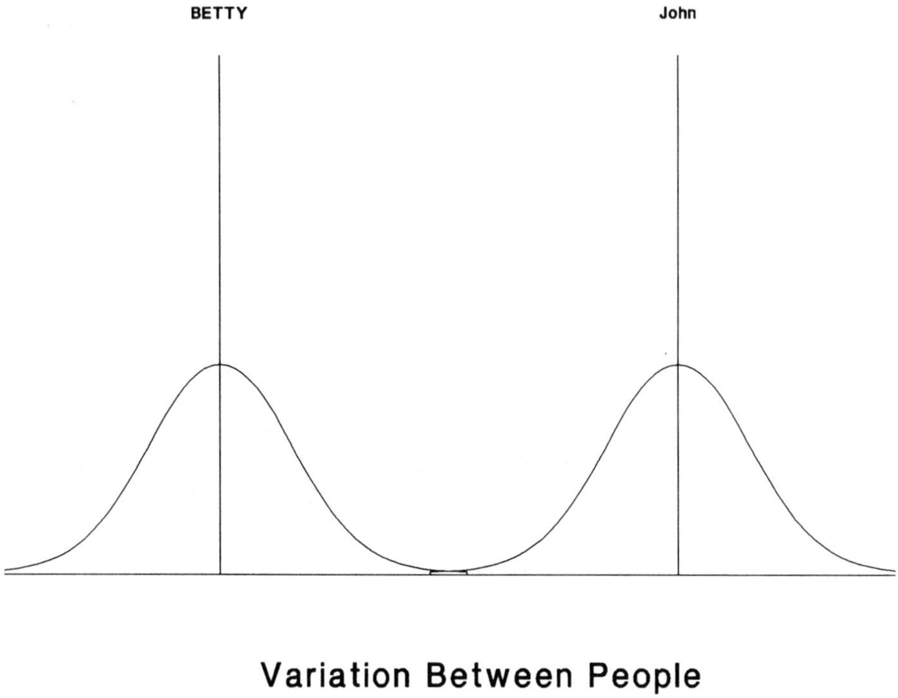

Figure 11.3 Reproducibility.

judgment on quality are trained to be able to classify correctly the quality characteristics of items. Often inspection standards, such as photographs or other visual aids, are provided at the point of inspection so that those involved may constantly refer to them and "recalibrate" themselves.

Evaluating Repeatability, Reproducibility, and Stability Errors Using Control Charts

The basic concept behind any control chart is simply that limits are computed which define the variation expected from common causes. Any variability beyond these limits is probably due to a special cause of variation. With this in mind, we can use control charts to investigate the potential sources of measurement error as special causes of variation.

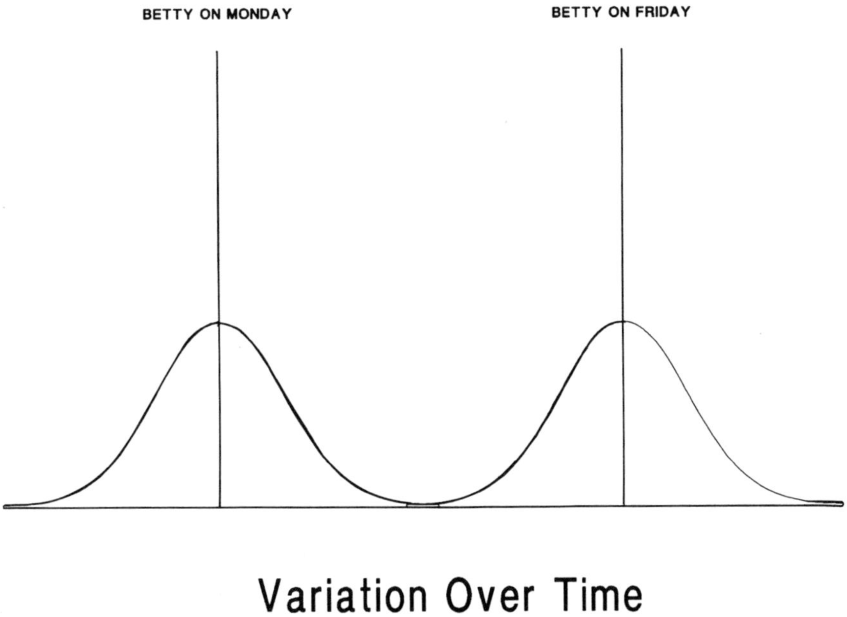

Figure 11.4 Stability.

This is done by carefully planning and organizing the data collection. One procedure is as follows:

1. Select five or more items for inspection (in the case of attributes, groups of items). As always, the more items used the more expensive your test, but the more accurate your estimates.
2. Have each inspector check each item at least twice in the same sitting. Assign the items to the inspector at random and take care to conceal his previous inspection results from him. Be sure that each inspector understands and follows the correct inspection procedure.
3. Plot a control chart using Figure 11.5 as a guide.

The important thing to note in Figure 11.5 is that the layout is such that more parts and inspectors can be added indefinitely. The "meas 1"

	INSPECTOR A			
	meas 1	meas 2	average	range
part no. 1				
part no. 2				
⋮	⋮	⋮	⋮	⋮

	INSPECTOR B			
	meas 1	meas 2	average	range
part no. 1				
part no. 2				
⋮	⋮	⋮	⋮	⋮

Figure 11.5 Measurement error control chart layout.

in the figure means measurement number 1. The range chart measures *repeatability*. If repeatability were perfect, every range would be zero, the average range would be zero, and the control limits would be zero, indicating a measurement system with perfect repeatability.

In the control chart layout shown, we can separate the results for individual parts and inspectors. By providing this sort of layout we are, in effect, comparing the part-to-part and inspector-to-inspector variability to the repeatability. Here are some guidelines for interpreting the results:

1. The range control chart should be in control. The smaller the average range, the better.
2. The ranges for all the inspectors should be the same (in control). If they not, the measurement system is not *reproducible*. Look for a special cause. Also, the averages for the different inspectors should follow the same pattern. However, because of the method used to get the control limits for the averages chart, that chart will not necessarily show statistical control (see item 4 below).
3. If the range chart is out of control for a certain part for all inspec-

tors, find out why that part is so hard to check. For example, perhaps you are checking the diameter of a hole that is out of round.

4. *The averages control chart should be out of control.* The limits for the averages control chart are based on the average range, and the average range is based on measurement repeatability. However, the part-to-part variation is determined by the process, and the process variation should be much greater than the measurement variation. Since the averages control chart shows the part-to-part variation, it should be out of control. If the averages chart is not out of control, the measurement system cannot tell one part from another and cannot be used for effective process control.
5. The pattern on the averages chart should match for each inspector. This can be investigated more easily if a separate chart is made for each inspector, using the same chart scales.

This entire study can be repeated at a later date. A direct comparison of the results of the two studies is a way of evaluating the *stability* of the measurement system. Even if only some of the study can be done again, it will provide a good indication of stability.

You will recall that if the range chart is in control, the standard deviation can be computed by dividing the average range by a constant. When subgroups of two are used, as they are here, sigma = R-bar/1.128. This sigma is based only on measurement error. You can also compute sigma using Equation (9) in Chapter 6. However, this second value of sigma combines both process variability and measurement error. These two values can be used to estimate the percentage of the total variation accounted for by repeatability error. The necessary equations are shown below

$$\sigma^2_{\text{total}} = \sigma^2_{\text{process}} + \sigma^2_{\text{measurement error}} \tag{49}$$

where σ for the total and measurement error is computed from the average range; σ^2 is just the value of σ squared.

$$\text{Error \%} = 100 \times \frac{\sigma^2_{\text{error}}}{\sigma^2_{\text{total}}} \tag{50}$$

$$\text{Process \%} = 100 - \text{error \%} \tag{51}$$

Example. The following data are from an actual measurement error study involving two inspectors. The inspectors were asked to inspect the surface finish of five parts. Each part was checked twice. The gage used is called a "profilometer," and it records the surface roughness in microinches. It has a digital readout that shows one decimal place.

Inspector A Data

Part	Meas 1	Meas 2	Average	Range
1	111.9	112.3	112.10	0.4
2	108.1	108.1	108.10	0.0
3	124.9	124.6	124.75	0.3
4	118.6	118.7	118.65	0.1
5	133.0	130.7	131.85	2.3

Inspector B Data

Part	Meas 1	Meas 2	Average	Range
1	111.4	112.9	112.15	1.5
2	107.7	108.4	108.05	0.7
3	124.6	124.2	124.40	0.4
4	120.0	119.3	119.65	0.7
5	130.4	130.1	130.25	0.3

From these data we get, using the procedures described in Chapter 7 on X-bar and range charts,

$$R\text{-bar} = 0.67$$

$$\text{UCL}_R = 3.267 \times R\text{-bar} = 2.1889$$

Comparing the ranges to the upper control limit, we see that the R chart is in control for all except part number 5, inspector A. The engineer who conducted the experiment had noted that for the first measurement the inspector had checked the part in a slightly different location than was specified. On rechecking, he found that that area was, in fact, different. Knowing this, we can drop this group, and the new value of R-bar is 0.49. With all the remaining groups in control (UCL = 1.6), we calculate the measurement error sigma as

$$\sigma_{\text{error}} = \frac{R\text{-bar}}{d_2} = \frac{0.49}{1.128} = 0.4344$$

and the error variance, σ^2, is just the square of the above number,

$$\sigma^2_{\text{error}} = (0.4344)^2 = 0.1887$$

The process standard deviation was calculated from Equation (11) given in Chapter 6. Part 5, inspector A, data were not included. This gives the following sigma total and variance total:

$$\sigma_{\text{total}} = 7.491$$

$$\sigma^2_{\text{total}} = 56.115$$

Thus, using Equations (50) and (51), the percentage measurement error and process error are

$$\text{Error }\% = \frac{0.1187}{56.115} \times 100 = 0.21\%$$

$$\text{Process }\% = 100 - 0.21 = 99.79\%$$

Clearly, this measurement system is sufficiently repeatable for this process. The control charts, presented in Figure 11.6, show quite clearly that the two inspectors tend to get the same results.Thus, the measurement system is reproducible. The entire experiment should be repeated at a future date and compared to these results to determine whether the measurement system is also stable.

SUMMARY

This chapter introduced the basic terminology of measurement error. A method of classifying measurement error was presented. Analysis of measurement error with control charts was described and demonstrated. A statistical method of separating process variation from measurement error was given.

Not covered was error in sensory testing or attribute measures. More sophisticated methods of analyzing measurement error are also available (e.g., designed experiments) but were not discussed. The effect of measurement error on producer's or consumer's risk was not

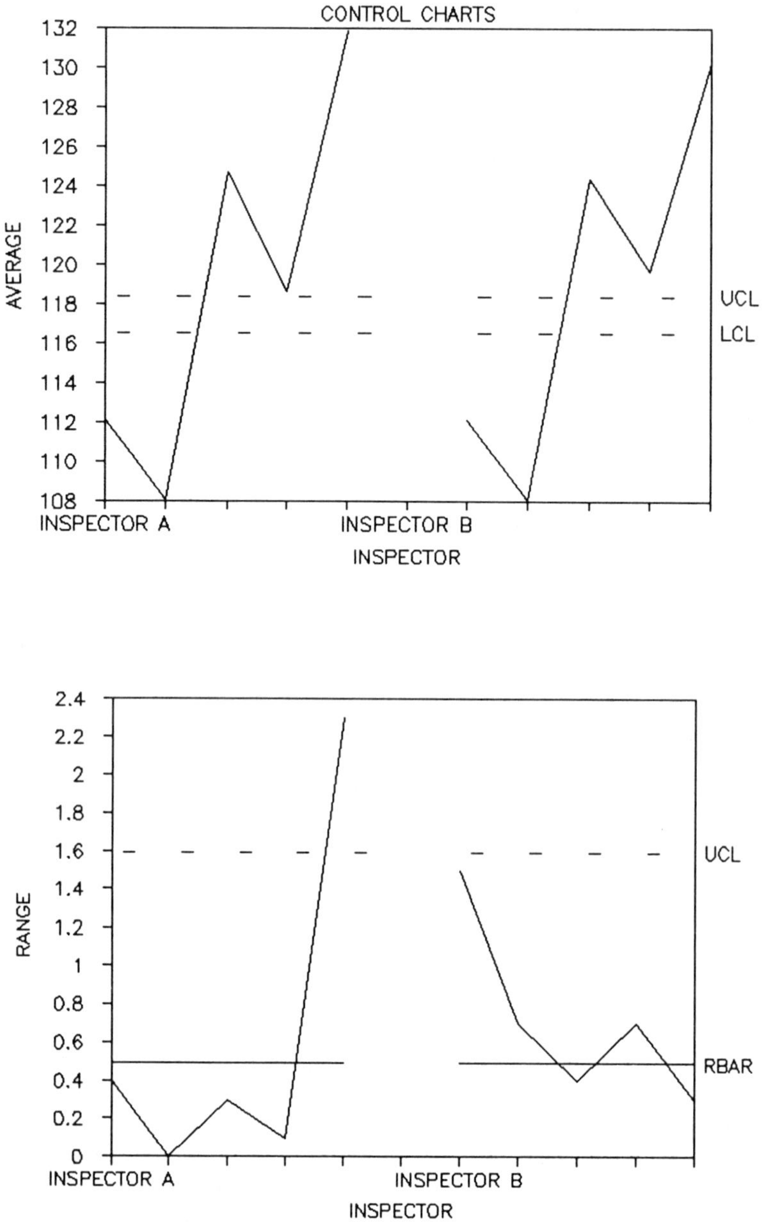

Figure 11.6 Measurement error control charts.

evaluated. Finally, the chapter omits descriptions of existing standards and specifications for calibration.

RECOMMENDED READING LIST

15–16, 19, 80.

12

Japanese Approach to Quality Management

INTRODUCTION

By now it is no secret that the Japanese are major-league competitors in the international marketplace. Equally well known is the fact that this position came about partly as a result of the superior quality of Japanese goods. Since this superiority encompasses a broad range of products, from microelectronic devices to heavy industrial machinery, it seems likely that Japanese manufacturers as a group are doing something noteworthy. This chapter explores the way in which the Japanese approach quality.

Perhaps a starting point is to examine history. It is important to remember that the Japanese were engaged in manufacturing in a significant way well before World War II, as evidenced by their arsenal of weapons. Their fleet of aircraft carriers and airplanes attest to that. However, the reality in postwar Japan was that their quality levels were too poor to give them much access to international markets. The American occupation forces understood the significance of export markets to Japan, which is only a small island with 110 million people.

Clearly, quality improvement had to be one of Japan's highest priorities.

Actually, the Japanese had begun their education somewhat before the American occupation. During World War II the Japanese were plagued by quality problems. Part of their wartime quality improvement involved translating works on statistical quality control methods into Japanese. However, the real push for implementing modern quality control methods came with a series of seminars conducted in 1950 by Dr. W. Edwards Deming, an American statistician and quality expert. Perhaps more important than the technical tools, Deming's philosophy of management was imparted to Japanese top management. The wholehearted endorsement of Deming's philosophy by Japanese top management was probably the single most important factor in Japan's turnaround in quality. With top management as champions, the adoption of Deming's philosophy was assured. The Japanese have honored Deming by naming their most prestigious industrial award in his honor: the Deming award.

Deming's philosophy has been embodied in his 14 Points for Management, which are listed in condensed form in Table 12.1. These 14 points are elaborated on in a number of different books; see Deming (1982), Scherkenbach (1986), and Gitlow and Gitlow (1987).

Other eminent American quality experts also made important contributions to Japanese quality improvement efforts. Our overall approach to quality management, including the use of statistical methods, was taught to eager Japanese audiences throughout the 1950s and 1960s. In addition to SPC, the Japanese learned of and adopted total quality control and zero defects programs. In return, we learned a great deal about quality improvement from the Japanese. Table 12.2 summarizes the exchange of information between America and Japan.

As Table 12.2 shows, the information exchange has been a two-way street. Many authors and consultants seem to believe that all useful QC methods are the property of American genius, expropriated unfairly by the Japanese. This is simply not the case. The free exchange of ideas and information has been a tradition in both American and Japan.

DIFFERENCES

As a quality professional, I have been interested in the Japanese for quite some time. The quality literature first began to mention Japan in the 1960s. By the mid-1970s Japan's high levels of quality began to draw

Table 12.1 Deming's 14 Points for Management

1. Innovate and allocate resources to fulfill the long-range needs of the company and customer rather than short-term profitability.
2. Discard the old philosophy of accepting defective products.
3. Eliminate dependence on mass inspection for quality control; instead, depend on process control through statistical methods.
4. Reduce the number of multiple source suppliers. Price has no meaning without an integral consideration for quality. Encourage suppliers to use statistical process control.
5. Use statistical techniques to identify the two sources of waste—system (85%) and local (15%) faults; strive to constantly reduce this waste.
6. Institute more thorough, better job-related training.
7. Provide supervision with knowledge of statistical methods; encourage use of these methods to identify which defects should be investigated for solution.
8. Reduce fear throughout the organization by encouraging open, two-way, nonpunitive communication.
9. Help reduce waste by encouraging design, research, and sales people to learn more about the problems of production.
10. Eliminate the use of goals and slogans to encourage productivity, unless training and management support is also provided.
11. Closely examine the impact of work standards. Do they consider quality or help anyone to do a better job? They often act as an impediment to quality.
12. Institute rudimentary statistical training on a broad scale.
13. Institute a vigorous program for retraining people in new skills, to keep up with changes in materials, methods, product designs, and machinery.
14. Make maximum use of statistical knowledge and talent in your company.

Table 12.2 American and Japanese Contributions to Quality Control

American contributions	Japanese contributions
Statistical QC methods (pre World War II)	New techniques: quality circles, fishbone diagrams, Taguchi methods
Deming's philosophy (1950)	Long-term orientation to management
Juran's approach to quality management (1954)	Externally focused management (customers, competitors, suppliers, government, society)
Feigenbaum's total quality control approach (1961)	Companywide involvement in QC

attention from outside the quality profession. When Japanese quality finally surpassed that of competing American goods, everyone took notice. Japanese superiority in quality and productivity became the subject of many studies. It was shown over and over again that Japanese goods were superior to their American counterparts. The list includes automobiles, videotapes, televisions, air conditioners, and a huge variety of other products.

A procession of American managers and consultants went to Japan to plumb the depths of this mystery. The result of this firsthand observation was the realization that there was nothing to see. The Japanese appeared to be using the same techniques as their American counterparts. Obviously, however, something is different. One Japanese quality expert, Dr. Kaoru Ishikawa, offered some of his observations on how the Japanese approach to quality differs from that of Western nations (Ishikawa, 1985):

1. Companywide quality control; participation by all members of the organization in quality control
2. Education and training in quality control
3. Quality control circle activities
4. Quality control audits (Deming prize audits and presidential audit)
5. Utilization of statistical methods
6. Nationwide quality control promotion activities

There are also differences in how American managers and Japannese managers define the essence of quality control. The American view is that quality has been achieved if the customer receives product that conforms to engineering requirements (i.e., zero defects). In fact, many Americans believe that an acceptable quality level (AQL) might still allow defective product to be delivered to the customer as a long-term practice. The Japanese view is that firms have a responsibility to society to reduce, and eventually eliminate, all negative impact on society caused by the firm's products and services. Quality, in the Japanese view, is defined by the customer, not by the producer.

MEASURING QUALITY

Efforts to measure quality are vital to the quality control effort. Much of this book has dealt with the subject of quality measurement. Since the importance of quality measurement is recognized by both American and Japanese firms, an examination of the way in which we measure quality might provide further clues to the differences in quality levels. Ishikawa contrasts the ways in which Japanese and Western firms express quality. He provides seven "pointers in expressing quality" (Ishikawa, 1985, pp. 49–55):

1. Determine the assurance unit (what we are to measure).
2. Determine the measuring method.
3. Determine the relative importance of quality characteristics.
4. Arrive at a consensus on defects and flaws.
5. Expose latent defects.
6. Observe quality statistically.
7. "Quality of design" and "quality of conformance."

From my experience, I believe that American manufacturers believe that they actually do all of these things. However, they are wrong. While all manufacturers may have done one or more of these things at one time or another, the application is so sporadic and haphazard that it can't be said that the *approach* was applied. The difference is crucial.

Another fundamental difference is the very way we look at quality. In America the way we think of quality is negative. Quality is measured as the absence of undesirable attributes (e.g., a unit "has quality" if it is nondefective). Ishikawa calls this *backward-looking quality*. The customer, in contrast, judges quality as the presence of desirable attributes. This is called *forward-looking quality*. The very use of the term *quality control* implies a backward-looking quality philosophy.

Even when we measure quality as the absence of undesirable attributes, we differ from the Japanese. Rework and repair are often so common that they are explicitly incorporated into manufacturing processes. On one occasion I heard a company vice-president accuse a quality improvement team of wasting their time trying to improve the quality of soldering on a circuit board, because "solder problems are never the cause of shipping problems." The team leader correctly pointed out that 40% of the cost of the circuit board was from solder touch-up and inspection after touch-up, a cost included as a normal part of the operation.

QUALITY CIRCLES

Quality circles are a Japanese contribution to quality technology. A quality circle is a small group that meets periodically to perform quality control activities. Circle membership is voluntary, and most circles are composed of people from the same work area. Circles are usually permanent and their activities are part of the broader companywide quality control effort.

Circle members are trained in the use of quality control techniques and supported by management. The idea is that all workers have an important contribution to make to the quality effort and the circle is one mechanism for allowing them the opportunity to make their contribution.

American business has adopted the quality circle approach on a massive scale, with mostly poor results. My experience has been that the circles are usually begun with great enthusiasm, produce a few significant results, and then die out. While there are many reasons for this, the most common ones are:

1. The quality circle in an American firm is isolated, not part of a companywide quality control effort. As a result, circles are usually unable to deal successfully with problems involving other areas of the company.
2. Key management personnel move about too frequently and circles are not provided with consistent leadership and management support.
3. Employees transfer in and out of circle work areas too frequently. Without stability in the membership, circles never develop into effective groups; building effective teams takes time.

VENDOR RELATIONS

The Japanese view vendors as an extension of their company. And well they should, since, on average, Japanese firms purchase 70% of the cost of their products from outside sources. Vendors are viewed as vital elements in the quality control program and are treated as such. Visits and other communications are frequent and routine. Japanese firms develop very close working relationships with a small number of suppliers. The smaller number means that each supplier can be given larger orders, which helps win their loyalty.

American companies also depend a great deal on their vendors, averaging 50% of their costs for purchased goods. However, we treat our vendors much differently than the Japanese. Multiple sources of supply are common; some government contracts even require them. Sources are pitted against each other and threatened with the loss of business. Communication is often one-sided, with the customer telling the vendor what the requirements are and informing the supplier when requirements are not satisfied. American suppliers are seldom brought in during the design phase, and their input at all stages is limited or nonexistent.

EMPLOYEE RELATIONS

I once attended an international quality conference and listened to a number of Japanese lectures. The speakers repeatedly spoke of "loving the workers" and of the workers "loving the company." In Japanese businesses the fundamental assumptions regarding employees are different than those in American firms. Many, if not most, Japanese place their loyalty to their employers above their families.

Most American businesses have paid lip service to employee involvement. Some have even invested a great deal of money in this area. The results have been mixed but for the most part unimpressive. Our history has created an atmosphere of mistrust and hostility between management and employees that is difficult to break down. Management and workers have been isolated from one another both culturally and physically. Many companies actually promote this with such things as separate office buildings, cafeterias, rest rooms, and parking lots. The solution will require more than slogans; management must demonstrate their sincerity in meaningful ways.

QUALITY AND MARKETING

The Japanese quality control effort includes aggressive pursuit of customer feedback. The marketing function plays an important role in this. This effort, combined with the forward-looking quality philosophy, actually drives the quality control effort. Customer feedback is used to assess the effectiveness of current quality control efforts and to provide the basis for improved process and product designs.

American quality control efforts make little use of marketing information, except perhaps to track warranty costs. Customer feedback is usually limited to complaints, and the complaints seldom lead to long-term corrective action. Redesign of processes or products based on customer feedback is rare.

JUST IN TIME

Many Americans have heard of just-in-time production, but few realize that it is intimately related to quality control. Just in time, or JIT, is a production goal. The objective is to operate without safety stock. The basic elements of JIT include cutting lot sizes, cutting setup times, and cutting purchase order costs. As used in Japan, JIT is part of a broader companywide effort that includes total quality control, employee involvement, and vendor involvement. Safety stock is viewed as a "narcotic drug" that hides symptoms that something is wrong and requires management attention.

American companies have been quick to adopt JIT. As often as not, it has been a disaster. The reason for this is that American companies fail to realize that JIT must be part of a companywide effort that includes total quality control, employee involvement, and vendor involvement. As stated above, safety stock is a narcotic drug that hides pain symptoms. We must complete the surgery before we remove the anesthesia!

PERFECTION VERSUS AQL

American quality control has long used a term called the acceptable quality level, or AQL. The term is defined as the maximum percentage defective that is acceptable as a long-term process average. This has become more than a mere term for many American companies: it has become an operating philosophy. "It meets the AQL" has become the battle cry for an entire generation of manufacturing managers fighting to maintain the status quo.

The Japanese philosophy is the antithesis of the AQL. It is, in a word, perfection. The concept of perfection goes beyond zero defects. It seeks as an ultimate goal the elimination of variation in quality. Perfection is, of course, unattainable. However, its value as a guiding philosophy is considerable, and its advantages over the AQL philosophy are obvious.

AMERICAN INNOVATORS

Probably the most successful American innovators to date have been the Japanese themselves! While some in this country continue to argue that the Japanese success is based on cultural differences, the fact is that Japanese management has already successfully transformed several companies in the United States. For example, a plant under American management was manufacturing television sets with a defect rate of 180 defects per 100 sets. The plant was purchased by a Japanese company. After three years of Japanese management the defect rate was reduced to between 3 and 4 defects per 100 sets. The improvement was made with the same work force and equipment. There are many similar stories.

A few American companies have managed to make dramatic improvements under American management. Two of the most noteworthy are Nashua Corporation of New Hampshire and Ford Motor Company. As a consultant to Ford and their suppliers, I have been involved with their program from the beginning. As a consultant to many other companies, I have also had the opportunity to note differences in the approach taken by others who haven't done as well. Ford and Nashua have realized, and the others have not that:

The change requires the *active* leadership of top management (as in the CEO and staff).

The leadership must be obsessed with quality: mere dabbling won't do.

Newer or better techniques are not the answer. Adopting JIT, or TQC, or quality circles, or even all of the techniques will accomplish little without a change in the underlying philosophy and management of the company.

The change must be total. The new philosophy isn't an addition to the current management system; it's a replacement.

The focus of the company must be external, not internal. The company must actively pursue input, including critical input, from customers, suppliers, and employees.

The Ford and Nashua programs are based on Deming's philosophy. Ford's experience is described by Bill Scherkenbach in his excellent book, which is listed in the bibliography (Scherkenbach, 1986).

SUMMARY

This chapter discussed the history of quality in Japan. Deming's 14 points for management were given. The major contributions of the American and Japanese to the quality field were considered. The major differences in the American and Japanese approaches to quality control were examined. A method of measuring or "expressing" quality was shown. Reasons why quality circles so often fail in American firms were provided. Different approaches to vendor and employee relations were offered. The "AQL fallacy" was exposed. Finally, the key ingredients of the successful American programs of quality improvement were provided.

The chapter did not provide a detailed road map for improving quality. In fact, although some maintain that it is possible to provide a generic road map for improving quality, this writer doubts that it is possible. However, the literature does offer a number of success stories from which the interested executive can glean a great deal of useful anecdotal information.

RECOMMENDED READING LIST

30, 32, 34–39

References and Recommended Reading

American Society for Quality Control. Generic Guidelines for Quality Systems. ANSI/ASQC Z1.15-1979. Milwaukee, Wis. 1979.

American Society for Quality Control. General Requirements for a Quality Program. ASQC Standard C1-1968. Milwaukee, Wis. 1968.

Brainard, E.H. Just How Good Are Vendor Surveys? *Quality Assurance*, pp. 22–25, August 1966.

Carlson, R.D., Gerber, J. and McHugh, J.F. *Manual of Quality Assurance Procedures and Forms*. Englewood Cliffs, N.J.: Prentice-Hall, 1981.

Crosby, P.B., *Quality Without Tears*. New York: Plume Books, 1984.

Crosby, P.B. *Quality Is Free*. New York: McGraw-Hill, 1979.

Deming, W.E. *Quality, Productivity, and Competitive Position*, Cambridge, Mass.: Center for Advanced Engineering Study, MIT, 1982.

Duncan, A.J. *Quality Control and Industrial Statistics*, 4th ed. Homewood, Ill.: Richard D. Irwin, 1974.

Feigenbaum, A.V. *Total Quality Control*, 3rd ed. New York: McGraw-Hill, 1983.

Field, D.L., ed. *Procurement Quality Control: A handbook of recommended practices*, 2nd ed. Milwaukee, Wis.: American Society for Quality Control, 1976.

Fukuda, R. *Managerial Engineering*, Stanford, Conn: Productivity Inc., 1983.

Gibra, I.N. Economically Optimal Determination of Parameters of np Control Charts, *Journal of Quality Technology*, vol. 10 no. 1, pp.12-19, 1978.

Gitlow, H.S., and Gitlow, S.J. *The Deming Guide to Quality and Competitive Position*. Englewood Cliffs, N.J.: Prentice-Hall, 1987.

Hayes, G.E., and Romig, H.G. *Modern Quality Control*, rev. ed. Encino, Calif.: Glencoe, 1982.

Ishikawa, K. *What Is Total Quality Control the Japanese Way*? Englewood Cliffs, N.J.: Prentice-Hall, 1985.

Juran, J.M., and Gryna, F.M. Jr. *Quality Planning and Analysis*, 2nd ed. New York: McGraw-Hill, 1980.

Juran, J.M., ed. *Quality Control Handbook*, 3rd ed. New York: McGraw-Hill, 1979.

Kivenko, K. *Quality Control for Management*. Englewood Cliffs, N.J.: Prentice-Hall, 1984.

Kackar, R. N. Off-Line Quality Control, Parameter Design, and the Taguchi Method, *Journal of Quality Technology,* vol. 17, no. 4, pp. 176-188, 1985.

McGreggor, D. *The Human Side of Enterprise*. New York: McGraw-Hill, 1960.

Montgomery, D. C. *Design and Analysis of Experiments*, 2nd ed. New York: Wiley, 1984.

Montogomery, D. C. *Introduction to Statistical Quality Control*, New York: Wiley, 1986.

Mood, A.M. On the Dependance of Sampling Inspection Plans Under Population Distributions, *Annals of Mathematical Statistics,* vol. 14, pp. 415–425, 1943.

Nelson, L.S. The Shewhart Control Chart—Tests for Special Causes, *Journal of Quality Technology,* vol. 16, no. 4, pp. 237–239, 1984.

Pyzdek, T. *An SPC Primer: Programmed Learning Guide to Statistical Process Control.* Tucson, Ariz.: Quality America, 1984.

Pyzdek, T. *Certified Quality Engineer Examination Study Guide.* Tucson, Ariz.: Quality America, 1986.

Pyzdek, T. The Impact of Quality Cost Reductions on Profits, *Quality Progress,* pp. 14–15, November 1976.

Scherkenbach, W.W. *The Deming Route to Quality and Productivity*. Washington, D.C.: Ceep Press, 1986.

Schilling, E.G. *Acceptance Sampling in Quality Control*. New York: Marcel Dekker, 1982.

Schonberger, R.J. *Japanese Manufacturing Techniques*. New York: Free Press, 1982.

Shewhart, W.A. *Economic Control of Quality of Manufactured Product.* New York: Van Nostrand, 1931. Reprinted by the American Society for Quality Control, Milwaukee Wis. 1980.

Taguchi, G. *Introduction to Quality Engineering.* White Plains, N.Y.: UNIPUB, 1986.

Taguchi, G. *System of Experimental Design.* White Plains, N.Y.: UNIPUB, 1987.

Thomas, D.W., et al. *Statistical Quality Control Handbook.* Indianapolis, Ind. AT&T, 1956.

Thorpe, J.F. *What Every Engineer Should Know About Product Liability.* New York: Marcel Dekker, 1979.

RECOMMENDED READING

The reading list given here is part of a list of materials recommended by ASQC for persons planning to take the examination to become certified quality engineers (CQEs). Since this book cannot cover the vast field of quality control in depth, these additional materials are required for anyone hoping to gain a comprehensive understanding of the subject matter. The reading list is grouped by subject, as done by ASQC. To facilitate further study, each chapter contains a list of recommended reading, referenced by the numbers given next to the entries in this list.

Quality Audit

1. ASQC Vendor-Vendee Technical Committee. *How to Conduct a Supplier Survey.* Milwaukee: ASQC, 1977.
2. ASQC Vendor-Vendee Technical Committee. *Procurement Quality Control,* 2nd ed. Milwaukee: ASQC, 1976.
3. Department of Defense. *Evaluations of Contractor's Quality Programs* (MIL-HBK-50). Naval Publications and Forms Center, 1965.
4. Department of Defense. *Evaluations of Contractor's Inspection Systems* (MIL-HBK-51). Naval Publications and Forms Center, 1967.
5. Ogen, J. E. *Product Quality Audit,* Booklet X7-2776/201. Autonetics Division, Rockwell International Corporation. Reprint. ASQC Education and Training Institute (ETI), 1973.

Quality Cost Analysis

6. ASQC Quality Cost-Cost Effectiveness Technical Committee. *Quality Costs—What and How,* 2nd ed. Milwaukee: ASQC, 1971.

7. ASQC Quality Cost-Cost Effectiveness Technical Committee. *Guide to Reducing Quality Costs*. Milwaukee: ASQC, 1977.

Quality Information Systems

8. ASQC Quality Cost-Cost Effectiveness Technical Committee. *Quality Costs—What and How*, 2nd ed. Milwaukee: ASQC, 1971.
9. ASQC Reliability Division. *Reliability Reporting Guide.* Milwaukee: ASQC, 1977.
10. Awad, Eliash M. *Business Data Processing,* 5th ed. Englewood Cliffs, N.J.: Prentice-Hall, 1980.
11. Davis, Godon B. *Introduction to Electronic Computers* 3rd ed. New York: McGraw-Hill, 1971.
12. Elliot, C.O., and Wasley, R.S. *Business Information Processing Systems,* 4th ed. Homewood Ill.: Richard D. Irwin, 1975.
13. Freund, John E., and Williams, Frank J. *Elementary Business Statistics; Modern Approach,* 4th ed. Englewood Cliffs, N.J.: Prentice-Hall, 1982.
14. Martin, E.W., Jr., and Perkins, W.C. *Computers and Information Systems, an Introduction.* Homewood, Ill.: Richard D. Irwin, 1973.

Metrology, Inspection, and Testing

15. American Society for Testing and Materials (ASTM). *Basic Principles of Sensory Evaluation,* Standard Technical Publication No. 433. ASTM, 1968.
16. ASTM. *Manual of Sensory Testing Methods,* Standard Technical Publication No. 434. ASTM, 1968.
17. American Society for Metals (ASM). *Nondestructive Inspection and Quality Control,* Vol. 2, *Metals Handbook*. ASTM, 1975.
18. ASQC Chemical Division. *Interlaboratory Testing Techniques*. Milwaukee: ASQC, 1978.
19. Department of Defense. *Evaluation of Contractors Calibration Systems* (MIL-HDBK-52). Naval Publications and Forms Center, 1968.
20. Hayward, Gordon P. *Introduction to Nondestructive Testing.* Inspector's Handbook Series. Milwaukee, ASQC, 1978.
21. Kennedy, C.W., and Andrews, D.E. *Inspection and Gaging,* 5th rev. ed. Industrial Press, 1977.
22. McGonnagle, Warren J. *Nondestructive Testing,* 2nd ed. New York: Gordon, 1971.

Quality Planning, Management, and Product Liability

23. ASQC. Specifications of General Requirements for a Quality Program, ANSI/ASQC C1-1978.
24. ASQC ETI. *Quality Control and Reliability Management* (Training Manual). Milwaukee: ASQC, 1969.

25. ASQC Energy Division. *Matrix of Nuclear Quality Assurance Program Requirements*. Milwaukee: ASQC, 1976.
26. ASQC Vendor-Vendee Technical Committee. *How to Conduct a Supplier Survey*. Milwaukee: ASQC, 1977.
27. ASQC Vendor-Vendee Technical Committee. *Procurement Quality Control*. Milwaukee: ASQC, 1976.
28. ASQC Vendor-Vendee Technical Committee. *How to Evaluate a Supplier's Product*. Milwaukee: ASQC, 1981.
29. ASQC Vendor-Vendee Technical Committee. *How to Establish Effective Quality Control for the Small Supplier*. Milwaukee: ASQC, 1981.
30. Caplan,Frank. *Quality System: A Sourcebook for Managers and Engineers*. Chilton, 1980.
31. Gary, Irwin, et al. *Product Liability: A Management Response*. American Management Association, 1975.
32. Juran, J.M. *Managerial Breakthrough*. New York: McGraw-Hill, 1964.
33. New Jersey Institute of Technology. *Proceedings, Product Liability Prevention Conferences*. Annals 1970–1977.

Human Factors and Motivation

34. Amsden, Davida M., and Amsden, Robert T. *QC Circles: Applications, Tools and Theory*. Milwaukee: ASQC, 1976.
35. ASQC Quality Motivation Technical Committee. *Quality Motivation Workbook*. Milwaukee: ASQC, 1978.
36. Crosby, Phillip B. *Quality Is Free*. New York: McGraw-Hill, 1979.
37. Herzberg, F. *Work and the Nature of Man*. New York: Crowell, 1973.
38. McCormick, Ernest J., and Ilgan, Daniel R. *Industrial Psychology*, 2nd ed. Englewood Cliffs, N.J.: Prentice-Hall, 1980.
39. McGregor, D. *Human Side of the Enterprise*. New York: McGraw-Hill, 1960.

Quality Control Principles

40. Charbonneau, Harvey C., and Webster, Gordon L. *Industrial Quality Control*. Englewood Cliffs, N.J.: Prentice-Hall, 1978.

Fundamentals of Probability, Statistical Quality Control, and Design of Experiments

41. Burr, Irving W. *Applied Statistical Methods*. Operations Research and Industrial Engineering Series. New York: Academic Press, 1973.
42. Chatfield, C. *Statistics for Technology: A Course in Applied Statistics*, 3rd ed. New York: Halsted, 1975.

43. Freund, John E. *Modern Elementary Statistics,* 6th ed. Englewood Cliffs, N.J.: Prentice-Hall, 1984.
44. Hine, J., and Wetherill. *A Programmed Text in Statistics.* New York: Halsted, 1975.
45. Lindgren, Bernard W., and McElrath, G.W. *Introduction to Probability and Statistics,* 4th ed. New York: Macmillan, 1978.
46. Rickmers, Albert D., and Todd, Hollis N. *Statistics: An Introduction.* New York: McGraw-Hill, 1967.
47. Spiegel, Murray R. *Statistics.* New York: McGraw-Hill, 1961.

Statistical Quality Control

48. ASQC. Definitions, Symbols, Formulas and Tables for Control Charts, ANSI/ASQC A1-1978.
49. ASQC. Terms and Symbols for Acceptance Sampling, ANSI/ASQC A2-1978.
50. ASQC. Quality Systems Terminology, ANSI/ASQC A3-1978.
51. ASQC. *Achieving Results Through Statistical Methods* (Home Study Course). Milwaukee: ASQC, 1977.
52. Burr, Irving W. *Statistical Quality Control Methods.* New York: Marcel Dekker, 1976.
53. Calvin, Thomas W. *How and When to Perform Bayesian Acceptance Sampling.* Milwaukee: ASQC, 1984.
54. Charbonneau, Harvey C. and Webster, Gordon L. *Industrial Quality Control.* Englewood Cliffs, N.J.: Prentice-Hall, 1978.
55. Department of Defense. *Sampling Procedures and Tables for Inspection by Variables* (MIL-STD-414). Naval Publications and Forms Center, 1968.
56. Department of Defense. *Sampling Procedures and Tables for Inspection by Attributes.* (MIL-STD-105). Naval Publications and Forms Center, 1964.
57. Dodge, Harold F., and Romig, Harry G. *Sampling Inspection Tables: Single and Double Sampling,* 2nd ed. New York: Wiley, 1959.
58. Duncan, Acheson J. *Quality Control and Industrial Statistics,* 4th ed. Homewood, Ill.: Richard D. Irwin, 1974.
59. Grant, Eugene L., and Leavenworth, Richard. *Statistical Quality Control,* 5th ed. New York: McGraw-Hill, 1980.
60. Schilling, Edward G. *Acceptance Sampling in Quality Control.* New York: Marcel Dekker, 1982.
61. Shapiro, Samuel S. *How to Test Normality and Other Distributional Assumptions.* Milwaukee: ASQC, 1980.
62. State University of Iowa Section, ASQC. *Quality Control Training Manual,* 2nd ed. ASQC, 1965.
63. Stephens, Kenneth S. *How to Perform Continuous Sampling (CSP)* Milwaukee: ASQC, 1979.

64. Stephens, Kenneth S. *How to Perform Skip-Lot and Chain Sampling.* Milwaukee: ASQC, 1982.

Experimental Design

65. Cornell, John A. *How to Run Mixture Experiments for Product Quality.* Milwaukee: ASQC, 1983.
66. Daves, O.L. *Design and Analysis of Industrial Experiments,* 2nd ed. New York: Longman, 1978.
67. Enrick, Norbert L., and Mottley, Harry E., Jr. *Manufacturing Improvement Through Experimentation.* General Instrument Corporation, 1968.
68. Fisher, Ronald A., and Prance, Ghiuean T. *Design of Experiments,* rev. 9th ed. New York: Hafner, 1966.
69. Hicks, C.R. *Fundamental Concepts in the Design of Experiments,* 2nd ed. New York: Holt, Rinehart & Winston, 1984.
70. Montgomery, Douglas C. *Design and Analysis of Experiments,* 2nd ed. New York: Wiley, 1984.

Reliability, Maintainability and Product Safety

71. ARINC Research Corporation. *Reliability Engineering.* Englewood Cliffs, N.J.: Prentice-Hall, 1964.
72. ASQC Reliability Division. *Reliability Reporting Guide.* Milwaukee: ASQC, 1977.
73. Calabro, S.R. *Reliability Principles and Practices.* New York: McGraw-Hill, 1962.
74. Goldman, A.S., and Slattery, T.B. *Maintainability: A Major Element of System Effectiveness,* 2nd ed. New York: Wiley, 1977.
75. Halpern, Siegmund. *Assurance Sciences: An Introduction to Quality and Reliability.* Englewood Cliffs, N.J.: Prentice-Hall, 1978.
76. Hammer, Willie. *Handbook of System and Product Safety.* Englewood Cliffs, N.J.: Prentice-Hall, 1972.
77. Ireson, William G. *Reliability Handbook.* New York: McGraw-Hill, 1966.
78. Lloyd, David K., and Lipow, Myron. *Reliability: Management Methods and Mathematics,* 2nd rev. ed. Lloyd and Lipow, 1977.
79. Nelson, Wayne. *Applied Life Data Analysis.* New York: Wiley, 1982.
80. Nelson, Wayne. *How to Analyze Data With Simple Plots.* Milwaukee: ASQC, 1979.
81. Nelson, Wayne. *How to Analyze Reliability Data.* Milwaukee: ASQC, 1983.

Appendix

Sampling Procedures and Tables for Inspection by Attributes

1. SCOPE

1.1 PURPOSE. This publication establishes sampling plans and procedures for inspection by attributes. When specified by the responsible authority, this publication shall be referenced in the specification, contract, inspection instructions, or other documents and the provisions set forth herein shall govern. The "responsible authority" shall be designated in one of the above documents.

1.2 APPLICATION. Sampling plans designated in this publication are applicable, but not limited, to inspection of the following:

a. End items.

b. Components and raw materials.

c. Operations.

d. Materials in process.

e. Supplies in storage.

f. Maintenance operations.

g. Data or records.

h. Administrative procedures.

These plans are intended primarily to be used for a continuing series of lots or batches. The plans may also be used for the inspection of isolated lots or batches, but, in this latter case, the user is cautioned to consult the operating characteristic curves to find a plan which will yield the desired protection (see 11.6).

1.3 INSPECTION. Inspection is the process of measuring, examining, testing, or otherwise comparing the unit of product (see 1.5) with the requirements.

1.4 INSPECTION BY ATTRIBUTES. Inspection by attributes is inspection whereby either the unit of product is classified simply as defective or nondefective, or the number of defects in the unit of product is counted, with respect to a given requirement or set of requirements.

1.5 UNIT OF PRODUCT. The unit of product is the thing inspected in order to determine its classification as defective or nondefective or to count the number of defects. It may be a single article, a pair, a set, a length, an area, an operation, a volume, a component of an end product, or the end product itself. The unit of product may or may not be the same as the unit of purchase, supply, production, or shipment.

This page and those through page 243 are reproduced by permission from the Department of Defense from Military Standard MIL-STD 105D, Sampling Procedures and Tables for Inspection by Attributes, 1963.

2. CLASSIFICATION OF DEFECTS AND DEFECTIVES

2.1 METHOD OF CLASSIFYING DEFECTS. A classification of defects is the enumeration of possible defects of the unit of product classified according to their seriousness. A defect is any nonconformance of the unit of product with specified requirements. Defects will normally be grouped into one or more of the following classes; however, defects may be grouped into other classes, or into subclasses within these classes.

2.1.1 CRITICAL DEFECT. A critical defect is a defect that judgment and experience indicate is likely to result in hazardous or unsafe conditions for individuals using, maintaining, or depending upon the product; or a defect that judgment and experience indicate is likely to prevent performance of the tactical function of a major end item such as a ship, aircraft, tank, missile or space vehicle. NOTE: For a special provision relating to critical defects, see 6.3.

2.1.2 MAJOR DEFECT. A major defect is a defect, other than critical, that is likely to result in failure, or to reduce materially the usability of the unit of product for its intended purpose.

2.1.3 MINOR DEFECT. A minor defect is a defect that is not likely to reduce materially the usability of the unit of product for its intended purpose, or is a departure from established standards having little bearing on the effective use or operation of the unit.

2.2 METHOD OF CLASSIFYING DEFECTIVES. A defective is a unit of product which contains one or more defects. Defectives will usually be classified as follows:

2.2.1 CRITICAL DEFECTIVE. A critical defective contains one or more critical defects and may also contain major and or minor defects. NOTE: For a special provision relating to critical defectives, see 6.3.

2.2.2 MAJOR DEFECTIVE. A major defective contains one or more major defects and may also contain minor defects but contains no critical defect.

2.2.3 MINOR DEFECTIVE. A minor defective contains one or more minor defects but contains no critical or major defect.

3. PERCENT DEFECTIVE AND DEFECTS PER HUNDRED UNITS

3.1 EXPRESSION OF NONCONFORMANCE. The extent of nonconformance of product shall be expressed either in terms of percent defective or in terms of defects per hundred units.

3.2 PERCENT DEFECTIVE. The percent defective of any given quantity of units of product is one hunderd times the number of defective units of product contained therein divided by the total number of units of product, i.e.:

$$\text{Percent defective} = \frac{\text{Number of defectives}}{\text{Number of units inspected}} \times 100$$

3.3 DEFECTS PER HUNDRED UNITS. The number of defects per hundred units of any given quantity of units of product is one hundred times the number of defects contained therein (one or more defects being possible in any unit of product) divided by the total number of units of product, i.e.:

$$\text{Defects per hundred units} = \frac{\text{Number of defects}}{\text{Number of units inspected}} \times 100$$

4. ACCEPTABLE QUALITY LEVEL (AQL)

4.1 USE. The AQL, together with the Sample Size Code Letter, is used for indexing the sampling plans provided herein.

4.2 DEFINITION. The AQL is the maximum percent defective (or the maximum number of defects per hundred units) that, for purposes of sampling inspection, can be considered satisfactory as a process average (see 11.2).

4.3 NOTE ON THE MEANING OF AQL. When a consumer designates some specific value of AQL for a certain defect or group of defects, he indicates to the supplier that his (the consumer's) acceptance sampling plan will accept the great majority of the lots or batches that the supplier submits, provided the process average level of percent defective (or defects per hundred units) in these lots or batches be no greater than the designated value of AQL. Thus, the AQL is a designated value of percent defective (or defects per hundred units) that the consumer indicates will be accepted most of the time by the acceptance sampling procedure to be used. The sampling plans provided herein are so arranged that the probability of acceptance at the designated AQL value depends upon the sample size, being generally higher for large samples than for small ones, for a given AQL. The AQL alone does not describe the protection to the consumer for individual lots or batches but more directly relates to what might be expected from a series of lots or batches, provided the steps indicated in this publication are taken. It is necessary to refer to the operating characteristic curve of the plan, to determine what protection the consumer will have.

4.4 LIMITATION. The designation of an AQL shall not imply that the supplier has the right to supply knowingly any defective unit of product.

4.5 SPECIFYING AQLs. The AQL to be used will be designated in the contract or by the responsible authority. Different AQLs may be designated for groups of defects considered collectively, or for individual defects. An AQL for a group of defects may be designated in addition to AQLs for individual defects, or subgroups, within that group. AQL values of 10.0 or less may be expressed either in percent defective or in defects per hundred units; those over 10.0 shall be expressed in defects per hundred units only.

4.6 PREFERRED AQLs. The values of AQLs given in these tables are known as preferred AQLs. If, for any product, an AQL be designated other than a preferred AQL, these tables are not applicable.

5. SUBMISSION OF PRODUCT

5.1 LOT OR BATCH. The term lot or batch shall mean "inspection lot" or "inspection batch," i.e., a collection of units of product from which a sample is to be drawn and inspected to determine conformance with the acceptability criteria, and may differ from a collection of units designated as a lot or batch for other purposes (e.g., production, shipment, etc.).

5.2 FORMATION OF LOTS OR BATCHES. The product shall be assembled into identifiable lots, sublots, batches, or in such other manner as may be prescribed (see 5.4). Each lot or batch shall, as far as is practicable,

5. SUBMISSION OF PRODUCT (Continued)

consist of units of product of a single type, grade, class, size, and composition, manufactured under essentially the same conditions, and at essentially the same time.

5.3 LOT OR BATCH SIZE. The lot or batch size is the number of units of product in a lot or batch.

5.4 PRESENTATION OF LOTS OR BATCHES. The formation of the lots or batches, lot or batch size, and the manner in which each lot or batch is to be presented and identified by the supplier shall be designated or approved by the responsible authority. As necessary, the supplier shall provide adequate and suitable storage space for each lot or batch, equipment needed for proper identification and presentation, and personnel for all handling of product required for drawing of samples.

6. ACCEPTANCE AND REJECTION

6.1 ACCEPTABILITY OF LOTS OR BATCHES. Acceptability of a lot or batch will be determined by the use of a sampling plan or plans associated with the designated AQL or AQLs.

6.2 DEFECTIVE UNITS. The right is reserved to reject any unit of product found defective during inspection whether that unit of product forms part of a sample or not, and whether the lot or batch as a whole is accepted or rejected. Rejected units may be repaired or corrected and resubmitted for inspection with the approval of, and in the manner specified by, the responsible authority.

6.3 SPECIAL RESERVATION FOR CRITICAL DEFECTS. The supplier may be required at the discretion of the responsible authority to inspect every unit of the lot or batch for critical defects. The right is reserved to inspect every unit submitted by the supplier for critical defects, and to reject the lot or batch immediately, when a critical defect is found. The right is reserved also to sample, for critical defects, every lot or batch submitted by the supplier and to reject any lot or batch if a sample drawn therefrom is found to contain one or more critical defects.

6.4 RESUBMITTED LOTS OR BATCHES. Lots or batches found unacceptable shall be resubmitted for reinspection only after all units are re-examined or retested and all defective units are removed or defects corrected. The responsible authority shall determine whether normal or tightened inspection shall be used, and whether reinspection shall include all types or classes of defects or for the particular types or classes of defects which caused initial rejection.

7. DRAWING OF SAMPLES

7.1 SAMPLE. A sample consists of one or more units of product drawn from a lot or batch, the units of the sample being selected at random without regard to their quality. The number of units of product in the sample is the sample size.

7.2 REPRESENTATIVE SAMPLING. When appropriate, the number of units in the sample shall be selected in proportion to the size of sublots or subbatches, or parts of the lot or batch, identified by some rational criterion.

7. DRAWING OF SAMPLES (Continued)

When representative sampling is used, the units from each part of the lot or batch shall be selected at random.

7.3 TIME OF SAMPLING. Samples may be drawn after all the units comprising the lot or batch have been assembled, or samples may be drawn during assembly of the lot or batch.

7.4 DOUBLE OR MULTIPLE SAMPLING. When double or multiple sampling is to be used, each sample shall be selected over the entire lot or batch.

8. NORMAL, TIGHTENED AND REDUCED INSPECTION

8.1 INITIATION OF INSPECTION. Normal inspection will be used at the start of inspection unless otherwise directed by the responsible authority.

8.2 CONTINUATION OF INSPECTION. Normal, tightened or reduced inspection shall continue unchanged for each class of defects or defectives on successive lots or batchs except where the switching procedures given below require change. The switching procedures given below require a change. The switching procedures shall be applied to each class of defects or defectives independently.

8.3 SWITCHING PROCEDURES.

8.3.1 NORMAL TO TIGHTENED. When normal inspection is in effect, tightened inspection shall be instituted when 2 out of 5 consecutive lots or batches have been rejected on original inspection (i.e., ignoring resubmitted lots or batches for this procedure).

8.3.2 TIGHTENED TO NORMAL. When tightened inspection is in effect, normal inspection shall be instituted when 5 consecutive lots or batches have been considered acceptable on original inspection.

8.3.3 NORMAL TO REDUCED. When normal inspection is in effect, reduced inspection shall be instituted providing that all of the following conditions are satisfied:

a. The preceding 10 lots or batches (or more, as indicated by the note to Table VIII) have been on normal inspection and none has been rejected on original inspection; and

b. The total number of defectives (or defects) in the samples from the preceding 10 lots or batches (or such other number as was used for condition "a" above) is equal to or less than the applicable number given in Table VIII. If double or multiple sampling is in use, all samples inspected should be included, not "first" samples only; and

c. Production is at a steady rate; and

d. Reduced inspection is considered desirable by the responsible authority.

8.3.4 REDUCED TO NORMAL. When reduced inspection is in effect, normal inspection shall be instituted if any of the following occur on original inspection:

a. A lot or batch is rejected; or

b. A lot or batch is considered acceptable under the procedures of 10.1.4; or

c. Production becomes irregular or delayed; or

d Other conditions warrant that normal inspection shall be instituted.

8.4 DISCONTINUATION OF INSPECTION. In the event that 10 consecutive lots or batches remain on tightened inspection (or such other number as may be designated by the responsible authority), inspection under the provisions of this document should be discontinued pending action to improve the quality of submitted material.

9. SAMPLING PLANS

9.1 SAMPLING PLAN. A sampling plan indicates the number of units of product from each lot or batch which are to be inspected (sample size or series of sample sizes) and the criteria for determining the acceptability of the lot or batch (acceptance and rejection numbers).

9.2 INSPECTION LEVEL. The inspection level determines the relationship between the lot or batch size and the sample size. The inspection level to be used for any particular requirement will be prescribed by the responsible authority. Three inspection levels: I, II, and III, are given in Table I for general use. Unless otherwise specified, Inspection Level II will be used. However, Inspection Level I may be specified when less discrimination is needed, or Level III may be specified for greater discrimination. Four additional special levels: S–1, S–2, S–3 and S–4, are given in the same table and may be used where relatively small sample sizes are necessary and large sampling risks can or must be tolerated.

NOTE: In the designation of inspection levels S–1 to S–4, care must be exercised to avoid AQLs inconsistent with these inspection levels.

9.3 CODE LETTERS. Sample sizes are designated by code letters. Table I shall be used to find the applicable code letter for the particular lot or batch size and the prescribed inspection level.

9.4 OBTAINING SAMPLING PLAN. The AQL and the code letter shall be used to obtain the sampling plan from Tables II, III or IV. When no sampling plan is available for a given combination of AQL and code letter, the tables direct the user to a different letter. The sample size to be used is given by the new code letter not by the original letter. If this procedure leads to different sample sizes for different classes of defects, the code letter corresponding to the largest sample size derived may be used for all classes of defects when designated or approved by the responsible authority. As an alternative to a single sampling plan with an acceptance number of 0, the plan with an acceptance number of 1 with its correspondingly larger sample size for a designated AQL (where available), may be used when designated or approved by the responsible authority.

9.5 TYPES OF SAMPLING PLANS. Three types of sampling plans: Single, Double and Multiple, are given in Tables II, III and IV, respectively. When several types of plans are available for a given AQL and code letter, any one may be used. A decision as to type of plan, either single, double, or multiple, when available for a given AQL and code letter, will usually be based upon the comparison between the administrative difficulty and the average sample sizes of the available plans. The average sample size of multiple plans is less than for double (except in the case corresponding to single acceptance number 1) and both of these are always less than a single sample size. Usually the administrative difficulty for single sampling and the cost per unit of the sample are less than for double or multiple.

10. DETERMINATION OF ACCEPTABILITY

10.1 PERCENT DEFECTIVE INSPECTION. To determine acceptability of a lot or batch under percent defective inspection, the applicable sampling plan shall be used in accordance with 10.1.1, 10.1.2, 10.1.3, 10.1.4, and 10.1.5.

10.1.1 SINGLE SAMPLING PLAN. The number of sample units inspected shall be equal to the sample size given by the plan. If the number of defectives found in the sample is equal to or less than the acceptance number, the lot or batch shall be considered acceptable. If the number of defectives is equal to or greater than the rejection number, the lot or batch shall be rejected.

10.1.2 DOUBLE SAMPLING PLAN. The number of sample units inspected shall be equal to the first sample size given by the plan. If the number of defectives found in the first sample is equal to or less than the first acceptance number, the lot or batch shall be considered acceptable. If the number of defectives found in the first sample is equal to or greater than the first rejection number, the lot or batch shall be rejected. If the number of defectives found in the first sample is between the first acceptance and rejection numbers, a second sample of the size given by the plan shall be inspected. The number of defectives found in the first and second samples shall be accumulated. If the cumulative number of defectives is equal to or less than the second acceptance number, the lot or batch shall be considered acceptable. If the cumulative number of defectives is equal to or greater than the second rejection number, the lot or batch shall be rejected.

10.1.3 MULTIPLE SAMPLE PLAN. Under multiple sampling, the procedure shall be similar to that specified in 10.1.2, except that the number of successive samples required to reach a decision may be more than two.

10.1.4 SPECIAL PROCEDURE FOR REDUCED INSPECTION. Under reduced inspection, the sampling procedure may terminate without either acceptance or rejection criteria having been met. In these circumstances, the lot or batch will be considered acceptable, but normal inspection will be reinstated starting with the next lot or batch (see 8.3.4 (b)).

10.2 DEFECTS PER HUNDRED UNITS INSPECTION. To determine the acceptability of a lot or batch under Defects per Hundred Units inspection, the procedure specified for Percent Defective inspection above shall be used, except that the word "defects" shall be substituted for "defectives."

11. SUPPLEMENTARY INFORMATION

11.1 OPERATING CHARACTERISTIC CURVES. The operating characteristic curves for normal inspection, shown in Table X (pages 30–62), indicate the percentage of lots or batches which may be expected to be accepted under the various sampling plans for a given process quality. The curves shown are for single sampling; curves for double and multiple sampling are matched as closely as practicable. The O. C. curves shown for AQLs greater than 10.0 are based on the Poisson distribution and are applicable for defects per hundred units inspection; those for AQLs of 10.0 or less and sample sizes of 80 or less are based on the binomial distribution and are applicable for percent defec-

11. SUPPLEMENTARY INFORMATION (Continued)

tive inspection; those for AQLs of 10.0 or less and sample sizes larger then 80 are based on the Poisson distribution and are applicable either for defects per hundred units inspection, or for percent defective inspection (the Poisson distribution being an adequate approximation to the binomial distribution under these conditions). Tabulated values, corresponding to selected values of probabilities of acceptance (P_a, in percent) are given for each of the curves shown, and, in addition, for tightened inspection, and for defects per hundred units for AQLs of 10.0 or less and sample sizes of 80 or less.

11.2 PROCESS AVERAGE. The process average is the average percent defective or average number of defects per hundred units (whichever is applicable) of product submitted by the supplier for original inspection. Original inspection is the first inspection of a particular quantity of product as distinguished from the inspection of product which has been resubmitted after prior rejection.

11.3 AVERAGE OUTGOING QUALITY (AOQ). The AOQ is the average quality of outgoing product including all accepted lots or batches, plus all rejected lots or batches after the rejected lots or batches have been effectively 100 percent inspected and all defectives replaced by nondefectives.

11.4 AVERAGE OUTGOING QUALITY LIMIT (AOQL). The AOQL is the maximum of the AOQs for all possible incoming qualities for a given acceptance sampling plan. AOQL values are given in Table V–A for each of the single sampling plans for normal inspection and in Table V–B for each of the single sampling plans for tightened inspection.

11.5 AVERAGE SAMPLE SIZE CURVES. Average sample size curves for double and multiple sampling are in Table IX. These show the average sample sizes which may be expected to occur under the various sampling plans for a given process quality. The curves assume no curtailment of inspection and are approximate to the extent that they are based upon the Poisson distribution, and that the sample sizes for double and multiple sampling are assumed to be 0.631n and 0.25n respectively, where n is the equivalent single sample size.

11.6 LIMITING QUALITY PROTECTION. The sampling plans and associated procedures given in this publication were designed for use where the units of product are produced in a continuing series of lots or batches over a period of time. However, if the lot or batch is of an isolated nature, it is desirable to limit the selection of sampling plans to those, associated with a designated AQL value, that provide not less than a specified limiting quality protection. Sampling plans for this purpose can be selected by choosing a Limiting Quality (LQ) and a consumer's risk to be associated with it. Tables VI and VII give values of LQ for the commonly used consumer's risks of 10 percent and 5 percent respectively. If a different value of consumer's risk is required, the O.C. curves and their tabulated values may be used. The concept of LQ may also be useful in specifying the AQL and Inspection Levels for a series of lots or batches, thus fixing minimum sample size where there is some reason for avoiding (with more than a given consumer's risk) more than a limiting proportion of defectives (or defects) in any single lot or batch.

Table I Sample Size Code Letters.

Lot or batch size	Special inspection levels				General inspection levels		
	S-1	S-2	S-3	S-4	I	II	III
2 to 8	A	A	A	A	A	A	B
9 to 15	A	A	A	A	A	B	C
16 to 25	A	A	B	B	B	C	D
26 to 50	A	B	B	C	C	D	E
51 to 90	B	B	C	C	C	E	F
91 to 150	B	B	C	D	D	F	G
151 to 280	B	C	D	E	E	G	H
281 to 500	B	C	D	E	F	H	J
501 to 1200	C	C	E	F	G	J	K
1201 to 3200	C	D	E	G	H	K	L
3201 to 10000	C	D	F	G	J	L	M
10001 to 35000	C	D	F	H	K	M	N
35001 to 150000	D	E	G	J	L	N	P
150001 to 500000	D	E	G	J	M	P	Q
500001 and over	D	E	H	K	N	Q	R

Table II Single Sampling Plans for Normal Inspection (Master Table).

Sample size code letter	Sample size	Acceptable Quality Levels (normal inspection) 0.010	0.015	0.025	0.040	0.065	0.10	0.15	0.25	0.40	0.65	1.0	1.5	2.5	4.0	6.5	10	15	25	40	65	100	150	250	400	650	1000
		Ac Re	Ac Re	Ac Re	Ac Re	Ac Re	Ac Re	Ac Re	Ac Re	Ac Re	Ac Re	Ac Re	Ac Re	Ac Re	Ac Re	Ac Re	Ac Re	Ac Re	Ac Re	Ac Re	Ac Re	Ac Re	Ac Re	Ac Re	Ac Re	Ac Re	Ac Re
A	2	↓	↓	↓	↓	↓	↓	↓	↓	↓	↓	↓	↓	↓	↓	0 1	↓	↓	1 2	2 3	3 4	5 6	7 8	10 11	14 15	21 22	30 31
B	3	↓	↓	↓	↓	↓	↓	↓	↓	↓	↓	↓	↓	↓	0 1	↑	↓	1 2	2 3	3 4	5 6	7 8	10 11	14 15	21 22	30 31	44 45
C	5	↓	↓	↓	↓	↓	↓	↓	↓	↓	↓	↓	↓	0 1	↑	↓	1 2	2 3	3 4	5 6	7 8	10 11	14 15	21 22	30 31	44 45	↑
D	8	↓	↓	↓	↓	↓	↓	↓	↓	↓	↓	↓	0 1	↑	↓	1 2	2 3	3 4	5 6	7 8	10 11	14 15	21 22	30 31	44 45	↑	↑
E	13	↓	↓	↓	↓	↓	↓	↓	↓	↓	↓	0 1	↑	↓	1 2	2 3	3 4	5 6	7 8	10 11	14 15	21 22	30 31	44 45	↑	↑	↑
F	20	↓	↓	↓	↓	↓	↓	↓	↓	↓	0 1	↑	↓	1 2	2 3	3 4	5 6	7 8	10 11	14 15	21 22	↑	↑	↑	↑	↑	↑
G	32	↓	↓	↓	↓	↓	↓	↓	↓	0 1	↑	↓	1 2	2 3	3 4	5 6	7 8	10 11	14 15	21 22	↑	↑	↑	↑	↑	↑	↑
H	50	↓	↓	↓	↓	↓	↓	↓	0 1	↑	↓	1 2	2 3	3 4	5 6	7 8	10 11	14 15	21 22	↑	↑	↑	↑	↑	↑	↑	↑
J	80	↓	↓	↓	↓	↓	↓	0 1	↑	↓	1 2	2 3	3 4	5 6	7 8	10 11	14 15	21 22	↑	↑	↑	↑	↑	↑	↑	↑	↑
K	125	↓	↓	↓	↓	↓	0 1	↑	↓	1 2	2 3	3 4	5 6	7 8	10 11	14 15	21 22	↑	↑	↑	↑	↑	↑	↑	↑	↑	↑
L	200	↓	↓	↓	↓	0 1	↑	↓	1 2	2 3	3 4	5 6	7 8	10 11	14 15	21 22	↑	↑	↑	↑	↑	↑	↑	↑	↑	↑	↑
M	315	↓	↓	↓	0 1	↑	↓	1 2	2 3	3 4	5 6	7 8	10 11	14 15	21 22	↑	↑	↑	↑	↑	↑	↑	↑	↑	↑	↑	↑
N	500	↓	↓	0 1	↑	↓	1 2	2 3	3 4	5 6	7 8	10 11	14 15	21 22	↑	↑	↑	↑	↑	↑	↑	↑	↑	↑	↑	↑	↑
P	800	↓	0 1	↑	↓	1 2	2 3	3 4	5 6	7 8	10 11	14 15	21 22	↑	↑	↑	↑	↑	↑	↑	↑	↑	↑	↑	↑	↑	↑
Q	1250	0 1	↑	↓	1 2	2 3	3 4	5 6	7 8	10 11	14 15	21 22	↑	↑	↑	↑	↑	↑	↑	↑	↑	↑	↑	↑	↑	↑	↑
R	2000	↑	↑	1 2	2 3	3 4	5 6	7 8	10 11	14 15	21 22	↑	↑	↑	↑	↑	↑	↑	↑	↑	↑	↑	↑	↑	↑	↑	↑

↓ = Use first sampling plan below arrow. If sample size equals, or exceeds, lot or batch size, do 100 percent inspection.
↑ = Use first sampling plan above arrow.
Ac = Acceptance number.
Re = Rejection number.

Table II-B Single Sampling Plans for Tightened Inspection (Master Table).

Sample size code letter	Sample size	Acceptable Quality Levels (tightened inspection)																									
		0.010	0.015	0.025	0.040	0.065	0.10	0.15	0.25	0.40	0.65	1.0	1.5	2.5	4.0	6.5	10	15	25	40	65	100	150	250	400	650	1000
		Ac Re	Ac Re	Ac Re	Ac Re	Ac Re	Ac Re	Ac Re	Ac Re	Ac Re	Ac Re	Ac Re	Ac Re	Ac Re	Ac Re	Ac Re	Ac Re	Ac Re	Ac Re	Ac Re	Ac Re	Ac Re	Ac Re	Ac Re	Ac Re	Ac Re	Ac Re
A	2	↓	↓	↓	↓	↓	↓	↓	↓	↓	↓	↓	↓	↓	↓	↓	↓	↓	↓	1 2	2 3	3 4	5 6	8 9	12 13	18 19	27 28
B	3	↓	↓	↓	↓	↓	↓	↓	↓	↓	↓	↓	↓	↓	↓	0 1	↓	↓	1 2	2 3	3 4	5 6	8 9	12 13	18 19	27 28	41 42
C	5	↓	↓	↓	↓	↓	↓	↓	↓	↓	↓	↓	↓	↓	0 1	↓	↓	1 2	2 3	3 4	5 6	8 9	12 13	18 19	27 28	41 42	↑
D	8	↓	↓	↓	↓	↓	↓	↓	↓	↓	↓	↓	↓	0 1	↓	↓	1 2	2 3	3 4	5 6	8 9	12 13	18 19	27 28	41 42	↑	↑
E	13	↓	↓	↓	↓	↓	↓	↓	↓	↓	↓	↓	0 1	↓	↓	1 2	2 3	3 4	5 6	8 9	12 13	18 19	27 28	41 42	↑	↑	↑
F	20	↓	↓	↓	↓	↓	↓	↓	↓	↓	↓	0 1	↓	↓	1 2	2 3	3 4	5 6	8 9	12 13	18 19	↑	↑	↑	↑	↑	↑
G	32	↓	↓	↓	↓	↓	↓	↓	↓	↓	0 1	↓	↓	1 2	2 3	3 4	5 6	8 9	12 13	18 19	↑	↑	↑	↑	↑	↑	↑
H	50	↓	↓	↓	↓	↓	↓	↓	↓	0 1	↓	↓	1 2	2 3	3 4	5 6	8 9	12 13	18 19	↑	↑	↑	↑	↑	↑	↑	↑
J	80	↓	↓	↓	↓	↓	↓	↓	0 1	↓	↓	1 2	2 3	3 4	5 6	8 9	12 13	18 19	↑	↑	↑	↑	↑	↑	↑	↑	↑
K	125	↓	↓	↓	↓	↓	↓	0 1	↓	↓	1 2	2 3	3 4	5 6	8 9	12 13	18 19	↑	↑	↑	↑	↑	↑	↑	↑	↑	↑
L	200	↓	↓	↓	↓	↓	0 1	↓	↓	1 2	2 3	3 4	5 6	8 9	12 13	18 19	↑	↑	↑	↑	↑	↑	↑	↑	↑	↑	↑
M	315	↓	↓	↓	↓	0 1	↓	↓	1 2	2 3	3 4	5 6	8 9	12 13	18 19	↑	↑	↑	↑	↑	↑	↑	↑	↑	↑	↑	↑
N	500	↓	↓	↓	0 1	↓	↓	1 2	2 3	3 4	5 6	8 9	12 13	18 19	↑	↑	↑	↑	↑	↑	↑	↑	↑	↑	↑	↑	↑
P	800	↓	↓	0 1	↓	↓	1 2	2 3	3 4	5 6	8 9	12 13	18 19	↑	↑	↑	↑	↑	↑	↑	↑	↑	↑	↑	↑	↑	↑
Q	1250	↓	0 1	↓	↓	1 2	2 3	3 4	5 6	8 9	12 13	18 19	↑	↑	↑	↑	↑	↑	↑	↑	↑	↑	↑	↑	↑	↑	↑
R	2000	0 1	↑	↓	1 2	2 3	3 4	5 6	8 9	12 13	18 19	↑	↑	↑	↑	↑	↑	↑	↑	↑	↑	↑	↑	↑	↑	↑	↑
S	3150			1 2																							

↓ = Use first sampling plan below arrow. If sample size equals or exceeds lot or batch size, do 100 percent inspection.
↑ = Use first sampling plan above arrow.
Ac = Acceptance number.
Re = Rejection number.

Table II-C Single Sampling Plans for Reduced Inspection (Master Table).

Sample size code letter	Sample size	Acceptable Quality Levels (reduced inspection)†																									
		0.010	0.015	0.025	0.040	0.065	0.10	0.15	0.25	0.40	0.65	1.0	1.5	2.5	4.0	6.5	10	15	25	40	65	100	150	250	400	650	1000
		Ac Re	Ac Re	Ac Re	Ac Re	Ac Re	Ac Re	Ac Re	Ac Re	Ac Re	Ac Re	Ac Re	Ac Re	Ac Re	Ac Re	Ac Re	Ac Re	Ac Re	Ac Re	Ac Re	Ac Re	Ac Re	Ac Re	Ac Re	Ac Re	Ac Re	Ac Re
A	2	↓	↓	↓	↓	↓	↓	↓	↓	↓	↓	↓	↓	↓	↓	0 1	↓	↓	1 2	2 3	3 4	5 6	7 8	10 11	14 15	21 22	30 31
B	2	↓	↓	↓	↓	↓	↓	↓	↓	↓	↓	↓	↓	↓	0 1	↑	↓	0 2	1 3	2 4	3 5	5 6	7 8	10 11	14 15	21 22	30 31
C	2	↓	↓	↓	↓	↓	↓	↓	↓	↓	↓	↓	↓	0 1	↑	↓	0 2	1 3	1 4	2 5	3 6	5 8	7 10	10 13	14 17	21 24	↑
D	3	↓	↓	↓	↓	↓	↓	↓	↓	↓	↓	↓	0 1	↑	↓	0 2	1 3	1 4	2 5	3 6	5 8	7 10	10 13	14 17	21 24	↑	↑
E	5	↓	↓	↓	↓	↓	↓	↓	↓	↓	↓	0 1	↑	↓	0 2	1 3	1 4	2 5	3 6	5 8	7 10	10 13	14 17	21 24	↑	↑	↑
F	8	↓	↓	↓	↓	↓	↓	↓	↓	↓	0 1	↑	↓	0 2	1 3	1 4	2 5	3 6	5 8	7 10	10 13	↑	↑	↑	↑	↑	↑
G	13	↓	↓	↓	↓	↓	↓	↓	↓	0 1	↑	↓	0 2	1 3	1 4	2 5	3 6	5 8	7 10	10 13	↑	↑	↑	↑	↑	↑	↑
H	20	↓	↓	↓	↓	↓	↓	↓	0 1	↑	↓	0 2	1 3	1 4	2 5	3 6	5 8	7 10	10 13	↑	↑	↑	↑	↑	↑	↑	↑
J	32	↓	↓	↓	↓	↓	↓	0 1	↑	↓	0 2	1 3	1 4	2 5	3 6	5 8	7 10	10 13	↑	↑	↑	↑	↑	↑	↑	↑	↑
K	50	↓	↓	↓	↓	↓	0 1	↑	↓	0 2	1 3	1 4	2 5	3 6	5 8	7 10	10 13	↑	↑	↑	↑	↑	↑	↑	↑	↑	↑
L	80	↓	↓	↓	↓	0 1	↑	↓	0 2	1 3	1 4	2 5	3 6	5 8	7 10	10 13	↑	↑	↑	↑	↑	↑	↑	↑	↑	↑	↑
M	125	↓	↓	↓	0 1	↑	↓	0 2	1 3	1 4	2 5	3 6	5 8	7 10	10 13	↑	↑	↑	↑	↑	↑	↑	↑	↑	↑	↑	↑
N	200	↓	↓	0 1	↑	↓	0 2	1 3	1 4	2 5	3 6	5 8	7 10	10 13	↑	↑	↑	↑	↑	↑	↑	↑	↑	↑	↑	↑	↑
P	315	↓	0 1	↑	↓	0 2	1 3	1 4	2 5	3 6	5 8	7 10	10 13	↑	↑	↑	↑	↑	↑	↑	↑	↑	↑	↑	↑	↑	↑
Q	500	0 1	↑	↓	0 2	1 3	1 4	2 5	3 6	5 8	7 10	10 13	↑	↑	↑	↑	↑	↑	↑	↑	↑	↑	↑	↑	↑	↑	↑
R	800	↑	↑	0 2	1 3	1 4	2 5	3 6	5 8	7 10	10 13	↑	↑	↑	↑	↑	↑	↑	↑	↑	↑	↑	↑	↑	↑	↑	↑

↓ = Use first sampling plan below arrow. If sample size equals or exceeds lot or batch size, do 100 percent inspection.
↑ = Use first sampling plan above arrow.
Ac = Acceptance number.
Re = Rejection number.
† = If the acceptance number has been exceeded, but the rejection number has not been reached, accept the lot, but reinstate normal inspection (see 10.1.4).

Table III-A Double Sampling Plans for Normal Inspection (Master Table).

Sample size code letter	Sample	Sample size	Cumulative sample size	Acceptable Quality Levels (normal inspection) 0.010	0.015	0.025	0.040	0.065	0.10	0.15	0.25	0.40	0.65	1.0	1.5	2.5	4.0	6.5	10	15	25	40	65	100	150	250	400	650	1000
				Ac Re	Ac Re	Ac Re	Ac Re	Ac Re	Ac Re	Ac Re	Ac Re	Ac Re	Ac Re	Ac Re	Ac Re	Ac Re	Ac Re	Ac Re	Ac Re	Ac Re	Ac Re	Ac Re	Ac Re	Ac Re	Ac Re	Ac Re	Ac Re	Ac Re	Ac Re
A				↓	↓	↓	↓	↓	↓	↓	↓	↓	↓	↓	↓	↓	↓	•	↓	↓	•	•	•	•	•	•	•	•	•
B	First	2	2	↓	↓	↓	↓	↓	↓	↓	↓	↓	↓	↓	↓	↓	•	↑	↓	0 2	0 3	1 4	2 5	3 7	5 9	7 11	11 16	17 22	25 31
	Second	2	4																	1 2	3 4	4 5	6 7	8 9	12 13	18 19	26 27	37 38	56 57
C	First	3	3	↓	↓	↓	↓	↓	↓	↓	↓	↓	↓	↓	↓	•	↑	↓	0 2	0 3	1 4	2 5	3 7	5 9	7 11	11 16	17 22	25 31	↑
	Second	3	6																1 2	3 4	4 5	6 7	8 9	12 13	18 19	26 27	37 38	56 57	
D	First	5	5	↓	↓	↓	↓	↓	↓	↓	↓	↓	↓	↓	•	↑	↓	0 2	0 3	1 4	2 5	3 7	5 9	7 11	11 16	17 22	25 31	↑	↑
	Second	5	10															1 2	3 4	4 5	6 7	8 9	12 13	18 19	26 27	37 38	56 57		
E	First	8	8	↓	↓	↓	↓	↓	↓	↓	↓	↓	↓	•	↑	↓	0 2	0 3	1 4	2 5	3 7	5 9	7 11	11 16	17 22	25 31	↑	↑	↑
	Second	8	16													1 2	3 4	4 5	6 7	8 9	12 13	18 19	26 27	37 38	56 57				
F	First	13	13	↓	↓	↓	↓	↓	↓	↓	↓	↓	•	↑	↓	0 2	0 3	1 4	2 5	3 7	5 9	7 11	11 16	↑	↑	↑	↑	↑	↑
	Second	13	26												1 2	3 4	4 5	6 7	8 9	12 13	18 19	26 27							
G	First	20	20	↓	↓	↓	↓	↓	↓	↓	↓	•	↑	↓	0 2	0 3	1 4	2 5	3 7	5 9	7 11	11 16	↑	↑	↑	↑	↑	↑	↑
	Second	20	40											1 2	3 4	4 5	6 7	8 9	12 13	18 19	26 27								
H	First	32	32	↓	↓	↓	↓	↓	↓	↓	•	↑	↓	0 2	0 3	1 4	2 5	3 7	5 9	7 11	11 16	↑	↑	↑	↑	↑	↑	↑	↑
	Second	32	64										1 2	3 4	4 5	6 7	8 9	12 13	18 19	26 27									
J	First	50	50	↓	↓	↓	↓	↓	↓	•	↑	↓	0 2	0 3	1 4	2 5	3 7	5 9	7 11	11 16	↑	↑	↑	↑	↑	↑	↑	↑	↑
	Second	50	100									1 2	3 4	4 5	6 7	8 9	12 13	18 19	26 27										
K	First	80	80	↓	↓	↓	↓	↓	•	↑	↓	0 2	0 3	1 4	2 5	3 7	5 9	7 11	11 16	↑	↑	↑	↑	↑	↑	↑	↑	↑	↑
	Second	80	160								1 2	3 4	4 5	6 7	8 9	12 13	18 19	26 27											
L	First	125	125	↓	↓	↓	↓	•	↑	↓	0 2	0 3	1 4	2 5	3 7	5 9	7 11	11 16	↑	↑	↑	↑	↑	↑	↑	↑	↑	↑	↑
	Second	125	250							1 2	3 4	4 5	6 7	8 9	12 13	18 19	26 27												
M	First	200	200	↓	↓	↓	•	↑	↓	0 2	0 3	1 4	2 5	3 7	5 9	7 11	11 16	↑	↑	↑	↑	↑	↑	↑	↑	↑	↑	↑	↑
	Second	200	400						1 2	3 4	4 5	6 7	8 9	12 13	18 19	26 27													
N	First	315	315	↓	↓	•	↑	↓	0 2	0 3	1 4	2 5	3 7	5 9	7 11	11 16	↑	↑	↑	↑	↑	↑	↑	↑	↑	↑	↑	↑	↑
	Second	315	630					1 2	3 4	4 5	6 7	8 9	12 13	18 19	26 27														
P	First	500	500	↓	•	↑	↓	0 2	0 3	1 4	2 5	3 7	5 9	7 11	11 16	↑	↑	↑	↑	↑	↑	↑	↑	↑	↑	↑	↑	↑	↑
	Second	500	1000					1 2	3 4	4 5	6 7	8 9	12 13	18 19	26 27														
Q	First	800	800	•	↑	↓	0 2	0 3	1 4	2 5	3 7	5 9	7 11	11 16	↑	↑	↑	↑	↑	↑	↑	↑	↑	↑	↑	↑	↑	↑	↑
	Second	800	1600				1 2	3 4	4 5	6 7	8 9	12 13	18 19	26 27															
R	First	1250	1250	↑	↑	0 2	0 3	1 4	2 5	3 7	5 9	7 11	11 16	↑	↑	↑	↑	↑	↑	↑	↑	↑	↑	↑	↑	↑	↑	↑	↑
	Second	1250	2500			1 2	3 4	4 5	6 7	8 9	12 13	18 19	26 27																

↓ = Use first sampling plan below arrow. If sample size equals or exceeds lot or batch size, do 100 percent inspection.
↑ = Use first sampling plan above arrow
Ac = Acceptance number
Re = Rejection number
• = Use corresponding single sampling plan (or alternatively, use double sampling plan below, where available).

Table III-B Double Sampling Plans for Tightened Inspection (Master Table).

Sample size code letter	Sample	Sample size	Cumulative sample size	Acceptable Quality Levels (tightened inspection)																									
				0.010	0.015	0.025	0.040	0.065	0.10	0.15	0.25	0.40	0.65	1.0	1.5	2.5	4.0	6.5	10	15	25	40	65	100	150	250	400	650	1000
				Ac Re	Ac Re	Ac Re	Ac Re	Ac Re	Ac Re	Ac Re	Ac Re	Ac Re	Ac Re	Ac Re	Ac Re	Ac Re	Ac Re	Ac Re	Ac Re	Ac Re	Ac Re	Ac Re	Ac Re	Ac Re	Ac Re	Ac Re	Ac Re	Ac Re	Ac Re
A				↓	↓	↓	↓	↓	↓	↓	↓	↓	↓	↓	↓	↓	↓	↓	↓	↓	↓	•	•	•	•	•	•	•	•
B	First	2	2	↓	↓	↓	↓	↓	↓	↓	↓	↓	↓	↓	↓	↓	↓	•	↓	↓	0 2	0 3	1 4	2 5	3 7	6 10	9 14	15 20	23 29
	Second	2	4																		1 2	3 4	4 5	6 7	11 12	15 16	23 24	34 35	52 53
C	First	3	3	↓	↓	↓	↓	↓	↓	↓	↓	↓	↓	↓	↓	↓	•	↓	↓	0 2	0 3	1 4	2 5	3 7	6 10	9 14	15 20	23 29	↑
	Second	3	6																	1 2	3 4	4 5	6 7	11 12	15 16	23 24	34 35	52 53	
D	First	5	5	↓	↓	↓	↓	↓	↓	↓	↓	↓	↓	↓	↓	•	↓	↓	0 2	0 3	1 4	2 5	3 7	6 10	9 14	15 20	23 29	↑	↑
	Second	5	10															1 2	3 4	4 5	6 7	11 12	15 16	23 24	34 35	52 53			
E	First	8	8	↓	↓	↓	↓	↓	↓	↓	↓	↓	↓	↓	•	↓	↓	0 2	0 3	1 4	2 5	3 7	6 10	9 14	15 20	23 29	↑	↑	↑
	Second	8	16														1 2	3 4	4 5	6 7	11 12	15 16	23 24	34 35	52 53				
F	First	13	13	↓	↓	↓	↓	↓	↓	↓	↓	↓	↓	•	↓	↓	0 2	0 3	1 4	2 5	3 7	6 10	9 14	↑	↑	↑	↑	↑	↑
	Second	13	26													1 2	3 4	4 5	6 7	11 12	15 16	23 24							
G	First	20	20	↓	↓	↓	↓	↓	↓	↓	↓	↓	•	↓	↓	0 2	0 3	1 4	2 5	3 7	6 10	9 14	↑	↑	↑	↑	↑	↑	↑
	Second	20	40												1 2	3 4	4 5	6 7	11 12	15 16	23 24								
H	First	32	32	↓	↓	↓	↓	↓	↓	↓	↓	•	↓	↓	0 2	0 3	1 4	2 5	3 7	6 10	9 14	↑	↑	↑	↑	↑	↑	↑	↑
	Second	32	64											1 2	3 4	4 5	6 7	11 12	15 16	23 24									
J	First	50	50	↓	↓	↓	↓	↓	↓	↓	•	↓	↓	0 2	0 3	1 4	2 5	3 7	6 10	9 14	↑	↑	↑	↑	↑	↑	↑	↑	↑
	Second	50	100										1 2	3 4	4 5	6 7	11 12	15 16	23 24										
K	First	80	80	↓	↓	↓	↓	↓	↓	•	↓	↓	0 2	0 3	1 4	2 5	3 7	6 10	9 14	↑	↑	↑	↑	↑	↑	↑	↑	↑	↑
	Second	80	160									1 2	3 4	4 5	6 7	11 12	15 16	23 24											
L	First	125	125	↓	↓	↓	↓	↓	•	↓	↓	0 2	0 3	1 4	2 5	3 7	6 10	9 14	↑	↑	↑	↑	↑	↑	↑	↑	↑	↑	↑
	Second	125	250								1 2	3 4	4 5	6 7	11 12	15 16	23 24												
M	First	200	200	↓	↓	↓	↓	•	↓	↓	0 2	0 3	1 4	2 5	3 7	6 10	9 14	↑	↑	↑	↑	↑	↑	↑	↑	↑	↑	↑	↑
	Second	200	400							1 2	3 4	4 5	6 7	11 12	15 16	23 24													
N	First	315	315	↓	↓	↓	•	↓	↓	0 2	0 3	1 4	2 5	3 7	6 10	9 14	↑	↑	↑	↑	↑	↑	↑	↑	↑	↑	↑	↑	↑
	Second	315	630							1 2	3 4	4 5	6 7	11 12	15 16	23 24													
P	First	500	500	↓	↓	•	↓	↓	0 2	0 3	1 4	2 5	3 7	6 10	9 14	↑	↑	↑	↑	↑	↑	↑	↑	↑	↑	↑	↑	↑	↑
	Second	500	1000						1 2	3 4	4 5	6 7	11 12	15 16	23 24														
Q	First	800	800	↓	•	↓	↓	0 2	0 3	1 4	2 5	3 7	6 10	9 14	↑	↑	↑	↑	↑	↑	↑	↑	↑	↑	↑	↑	↑	↑	↑
	Second	800	1600					1 2	3 4	4 5	6 7	11 12	15 16	23 24															
R	First	1250	1250	•	↑	↓	0 2	0 3	1 4	2 5	3 7	6 10	9 14	↑	↑	↑	↑	↑	↑	↑	↑	↑	↑	↑	↑	↑	↑	↑	↑
	Second	1250	2500				1 2	3 4	4 5	6 7	11 12	15 16	23 24																
S	First	2000	2000			0 2																							
	Second	2000	4000			1 2																							

⇩ = Use first sampling plan below arrow. If sample size equals or exceeds lot or batch size, do 100 percent inspection.

⇧ = Use first sampling plan above arrow.

Ac = Acceptance number

Re = Rejection number

• = Use corresponding single sampling plan (or, alternatively, use double sampling plan below, where available).

Table III-C Double Sampling Plans for Reduced Inspection (Master Table).

Sample size code letter	Sample	Sample size	Cumulative sample size	0.010 Ac Re	0.015 Ac Re	0.025 Ac Re	0.040 Ac Re	0.065 Ac Re	0.10 Ac Re	0.15 Ac Re	0.25 Ac Re	0.40 Ac Re	0.65 Ac Re	1.0 Ac Re	1.5 Ac Re	2.5 Ac Re	4.0 Ac Re	6.5 Ac Re	10 Ac Re	15 Ac Re	25 Ac Re	40 Ac Re	65 Ac Re	100 Ac Re	150 Ac Re	250 Ac Re	400 Ac Re	650 Ac Re	1000 Ac Re
				Acceptable Quality Levels (reduced inspection)†																									
A				↓	↓	↓	↓	↓	↓	↓	↓	↓	↓	↓	↓	↓	↓	•	↓	↓	•	•	•	•	•	•	•	•	•
B																	•	↑		•	•	•	•	•	•	•	•	•	•
C																•	↑	↓	•	•	•	•	•	•	•	•	•	•	↑
D	First	2	2	↓	↓	↓	↓	↓	↓	↓	↓	↓	↓	↓	•	↑	↓	0 2	0 3	0 4	0 4	1 5	2 7	3 8	5 10	7 12	11 17	↑	↑
	Second	2	4															0 2	0 4	1 5	3 6	4 7	6 9	8 12	12 16	18 22	26 30		
E	First	3	3	↓	↓	↓	↓	↓	↓	↓	↓	↓	↓	•	↑	↓	0 2	0 3	0 4	0 4	1 5	2 7	3 8	5 10	7 12	11 17	↑	↑	↑
	Second	3	6														0 2	0 4	1 5	3 6	4 7	6 9	8 12	12 16	18 22	26 30			
F	First	5	5	↓	↓	↓	↓	↓	↓	↓	↓	↓	•	↑	↓	0 2	0 3	0 4	0 4	1 5	2 7	3 8	5 10	↑	↑	↑	↑	↑	↑
	Second	5	10													0 2	0 4	1 5	3 6	4 7	6 9	8 12	12 16						
G	First	8	8	↓	↓	↓	↓	↓	↓	↓	↓	•	↑	↓	0 2	0 3	0 4	0 4	1 5	2 7	3 8	5 10	↑	↑	↑	↑	↑	↑	↑
	Second	8	16												0 2	0 4	1 5	3 6	4 7	6 9	8 12	12 16							
H	First	13	13	↓	↓	↓	↓	↓	↓	↓	•	↑	↓	0 2	0 3	0 4	0 4	1 5	2 7	3 8	5 10	↑	↑	↑	↑	↑	↑	↑	↑
	Second	13	26											0 2	0 4	1 5	3 6	4 7	6 9	8 12	12 16								
J	First	20	20	↓	↓	↓	↓	↓	↓	•	↑	↓	0 2	0 3	0 4	0 4	1 5	2 7	3 8	5 10	↑	↑	↑	↑	↑	↑	↑	↑	↑
	Second	20	40										0 2	0 4	1 5	3 6	4 7	6 9	8 12	12 16									
K	First	32	32	↓	↓	↓	↓	↓	•	↑	↓	0 2	0 3	0 4	0 4	1 5	2 7	3 8	5 10	↑	↑	↑	↑	↑	↑	↑	↑	↑	↑
	Second	32	64									0 2	0 4	1 5	3 6	4 7	6 9	8 12	12 16										
L	First	50	50	↓	↓	↓	↓	•	↑	↓	0 2	0 3	0 4	0 4	1 5	2 7	3 8	5 10	↑	↑	↑	↑	↑	↑	↑	↑	↑	↑	↑
	Second	50	100								0 2	0 4	1 5	3 6	4 7	6 9	8 12	12 16											
M	First	80	80	↓	↓	↓	•	↑	↓	0 2	0 3	0 4	0 4	1 5	2 7	3 8	5 10	↑	↑	↑	↑	↑	↑	↑	↑	↑	↑	↑	↑
	Second	80	160							0 2	0 4	1 5	3 6	4 7	6 9	8 12	12 16												
N	First	125	125	↓	↓	•	↑	↓	0 2	0 3	0 4	0 4	1 5	2 7	3 8	5 10	↑	↑	↑	↑	↑	↑	↑	↑	↑	↑	↑	↑	↑
	Second	125	250						0 2	0 4	1 5	3 6	4 7	6 9	8 12	12 16													
P	First	200	200	↓	•	↑	↓	0 2	0 3	0 4	0 4	1 5	2 7	3 8	5 10	↑	↑	↑	↑	↑	↑	↑	↑	↑	↑	↑	↑	↑	↑
	Second	200	400					0 2	0 4	1 5	3 6	4 7	6 9	8 12	12 16														
Q	First	315	315	•	↑	↓	0 2	0 3	0 4	0 4	1 5	2 7	3 8	5 10	↑	↑	↑	↑	↑	↑	↑	↑	↑	↑	↑	↑	↑	↑	↑
	Second	315	630				0 2	0 4	1 5	3 6	4 7	6 9	8 12	12 16															
R	First	500	500	↑	↑	0 2	0 3	0 4	0 4	1 5	2 7	3 8	5 10	↑	↑	↑	↑	↑	↑	↑	↑	↑	↑	↑	↑	↑	↑	↑	↑
	Second	500	1000			0 2	0 4	1 5	3 6	4 7	6 9	8 12	12 16																

↓ = Use first sampling plan below arrow. If sample size equals or exceeds lot or batch size, do 100 percent inspection.
↑ = Use first sampling plan above arrow.
Ac = Acceptance number.
Re = Rejection number.
• = Use corresponding single sampling plan (or alternatively, use double sampling plan below, when available.)
† = If, after the second sample, the acceptance number has been exceeded, but the rejection number has not been reached, accept the lot, but reinstate normal inspection (see 10.14).

Table IV-A Multiple Sampling Plans for Normal Inspection (Master Table).

Sample size code letter	Sample	Sample size	Cumulative sample size	Acceptable Quality Levels (normal inspection)																									
				0.010	0.015	0.025	0.040	0.065	0.10	0.15	0.25	0.40	0.65	1.0	1.5	2.5	4.0	6.5	10	15	25	40	65	100	150	250	400	650	1000
				Ac Re	Ac Re	Ac Re	Ac Re	Ac Re	Ac Re	Ac Re	Ac Re	Ac Re	Ac Re	Ac Re	Ac Re	Ac Re	Ac Re	Ac Re	Ac Re	Ac Re	Ac Re	Ac Re	Ac Re	Ac Re	Ac Re	Ac Re	Ac Re	Ac Re	Ac Re
A				↓	↓	↓	↓	↓	↓	↓	↓	↓	↓	↓	↓	↓	↓	•	↓	↓	•	•	•	•	•	•	•	•	•
B				↓	↓	↓	↓	↓	↓	↓	↓	↓	↓	↓	↓	↓	•	↑	↓	++	++	++	++	++	++	++	++	++	++
C				↓	↓	↓	↓	↓	↓	↓	↓	↓	↓	↓	↓	•	↑	↓	++	++	++	++	++	++	++	++	++	++	↑
D	First	2	2	↓	↓	↓	↓	↓	↓	↓	↓	↓	↓	↓	•	↑	↓	• 2	• 2	• 3	• 4	0 4	0 5	1 7	2 9	4 12	6 16	↑	↑
	Second	2	4															• 2	0 3	0 3	1 5	1 6	3 8	4 10	7 14	11 19	17 27		
	Third	2	6															0 2	0 3	1 4	2 6	3 8	6 10	8 13	13 19	19 27	29 39		
	Fourth	2	8															0 3	1 4	2 5	3 7	5 10	8 13	12 17	19 25	27 34	40 49		
	Fifth	2	10															1 3	2 4	3 6	5 8	7 11	11 15	17 20	25 29	36 40	53 58		
	Sixth	2	12															1 3	3 5	4 6	7 9	10 12	14 17	21 23	31 33	45 47	65 68		
	Seventh	2	14															2 3	4 5	6 7	9 10	13 14	18 19	25 26	37 38	53 54	77 78		
E	First	3	3	↓	↓	↓	↓	↓	↓	↓	↓	↓	↓	•	↑	↓	• 2	• 2	• 3	• 4	0 4	0 5	1 7	2 9	4 12	6 16	↑	↑	↑
	Second	3	6														• 2	0 3	0 3	1 5	1 6	3 8	4 10	7 14	11 19	17 27			
	Third	3	9														0 2	0 3	1 4	2 6	3 8	6 10	8 13	13 19	19 27	29 39			
	Fourth	3	12														0 3	1 4	2 5	3 7	5 10	8 13	12 17	19 25	27 34	40 49			
	Fifth	3	15														1 3	2 4	3 6	5 8	7 11	11 15	17 20	25 29	36 40	53 58			
	Sixth	3	18														1 3	3 5	4 6	7 9	10 12	14 17	21 23	31 33	45 47	65 68			
	Seventh	3	21														2 3	4 5	6 7	9 10	13 14	18 19	25 26	37 38	53 54	77 78			
F	First	5	5	↓	↓	↓	↓	↓	↓	↓	↓	↓	•	↑	↓	• 2	• 2	• 3	• 4	0 4	0 5	1 7	2 9	↑	↑	↑	↑	↑	↑
	Second	5	10													• 2	0 3	0 3	1 5	1 6	3 8	4 10	7 14						
	Third	5	15													0 2	0 3	1 4	2 6	3 8	6 10	8 13	13 19						
	Fourth	5	20													0 3	1 4	2 5	3 7	5 10	8 13	12 17	19 25						
	Fifth	5	25													1 3	2 4	3 6	5 8	7 11	11 15	17 20	25 29						
	Sixth	5	30													1 3	3 5	4 6	7 9	10 12	14 17	21 23	31 33						
	Seventh	5	35													2 3	4 5	6 7	9 10	13 14	18 19	25 26	37 38						
G	First	8	8	↓	↓	↓	↓	↓	↓	↓	↓	•	↑	↓	• 2	• 2	• 3	• 4	0 4	0 5	1 7	2 9	↑	↑	↑	↑	↑	↑	↑
	Second	8	16												• 2	0 3	0 3	1 5	1 6	3 8	4 10	7 14							
	Third	8	24												0 2	0 3	1 4	2 6	3 8	6 10	8 13	13 19							
	Fourth	8	32												0 3	1 4	2 5	3 7	5 10	8 13	12 17	19 25							
	Fifth	8	40												1 3	2 4	3 6	5 8	7 11	11 15	17 20	25 29							
	Sixth	8	48												1 3	3 5	4 6	7 9	10 12	14 17	21 23	31 33							
	Seventh	8	56												2 3	4 5	6 7	9 10	13 14	18 19	25 26	37 38							
H	First	13	13	↓	↓	↓	↓	↓	↓	↓	•	↑	↓	• 2	• 2	• 3	• 4	0 4	0 5	1 7	2 9	↑	↑	↑	↑	↑	↑	↑	↑
	Second	13	26											• 2	0 3	0 3	1 5	1 6	3 8	4 10	7 14								
	Third	13	39											0 2	0 3	1 4	2 6	3 8	6 10	8 13	13 19								
	Fourth	13	52											0 3	1 4	2 5	3 7	5 10	8 13	12 17	19 25								
	Fifth	13	65											1 3	2 4	3 6	5 8	7 11	11 15	17 20	25 29								
	Sixth	13	78											1 3	3 5	4 6	7 9	10 12	14 17	21 23	31 33								
	Seventh	13	91											2 3	4 5	6 7	9 10	13 14	18 19	25 26	37 38								
J	First	20	20	↓	↓	↓	↓	↓	↓	•	↑	↓	• 2	• 2	• 3	• 4	0 4	0 5	1 7	2 9	↑	↑	↑	↑	↑	↑	↑	↑	↑
	Second	20	40										• 2	0 3	0 3	1 5	1 6	3 8	4 10	7 14									
	Third	20	60										0 2	0 3	1 4	2 6	3 8	6 10	8 13	13 19									
	Fourth	20	80										0 3	1 4	2 5	3 7	5 10	8 13	12 17	19 25									
	Fifth	20	100										1 3	2 4	3 6	5 8	7 11	11 15	17 20	25 29									
	Sixth	20	120										1 3	3 5	4 6	7 9	10 12	14 17	21 23	31 33									
	Seventh	20	140										2 3	4 5	6 7	9 10	13 14	18 19	25 26	37 38									

Table IV-A (continued).

Sample size code letter	Sample	Sample size	Cumulative sample size	Acceptable Quality Levels (normal inspection)																									
				0.010	0.015	0.025	0.040	0.065	0.10	0.15	0.25	0.40	0.65	1.0	1.5	2.5	4.0	6.5	10	15	25	40	65	100	150	250	400	650	1000
				Ac Re	Ac Re	Ac Re	Ac Re	Ac Re	Ac Re	Ac Re	Ac Re	Ac Re	Ac Re	Ac Re	Ac Re	Ac Re	Ac Re	Ac Re	Ac Re	Ac Re	Ac Re	Ac Re	Ac Re	Ac Re	Ac Re	Ac Re	Ac Re	Ac Re	Ac Re
K	First	32	32	↓	↓	↓	↓	↓	•	↑	↓	# 2	# 2	# 3	# 4	0 4	0 5	1 7	2 9	↑	↑	↑	↑	↑	↑	↑	↑	↑	↑
	Second	32	64									# 2	0 3	0 3	1 5	1 6	3 8	4 10	7 14										
	Third	32	96									0 2	0 3	1 4	2 6	3 8	6 10	8 13	13 19										
	Fourth	32	128									0 3	1 4	2 5	3 7	5 10	8 13	12 17	19 25										
	Fifth	32	160									1 3	2 4	3 6	5 8	7 11	11 15	17 20	25 29										
	Sixth	32	192									1 3	3 5	4 6	7 9	10 12	14 17	21 23	31 33										
	Seventh	32	224									2 3	4 5	6 7	9 10	13 14	18 19	25 26	37 38										
L	First	50	50					•	↑	↓	# 2	# 2	# 3	# 4	0 4	0 5	1 7	2 9	↑										
	Second	50	100								# 2	0 3	0 3	1 5	1 6	3 8	4 10	7 14											
	Third	50	150								0 2	0 3	1 4	2 6	3 8	6 10	8 13	13 19											
	Fourth	50	200								0 3	1 4	2 5	3 7	5 10	8 13	12 17	19 25											
	Fifth	50	250								1 3	2 4	3 6	5 8	7 11	11 15	17 20	25 29											
	Sixth	50	300								1 3	3 5	4 6	7 9	10 12	14 17	21 23	31 33											
	Seventh	50	350								2 3	4 5	6 7	9 10	13 14	18 19	25 26	37 38											
M	First	80	80				•	↑	↓	# 2	# 2	# 3	# 4	0 4	0 5	1 7	2 9	↑											
	Second	80	160							# 2	0 3	0 3	1 5	1 6	3 8	4 10	7 14												
	Third	80	240							0 2	0 3	1 4	2 6	3 8	6 10	8 13	13 19												
	Fourth	80	320							0 3	1 4	2 5	3 7	5 10	8 13	12 17	19 25												
	Fifth	80	400							1 3	2 4	3 6	5 8	7 11	11 15	17 20	25 29												
	Sixth	80	480							1 3	3 5	4 6	7 9	10 12	14 17	21 23	31 33												
	Seventh	80	560							2 3	4 5	6 7	9 10	13 14	18 19	25 26	37 38												
N	First	125	125			•	↑	↓	# 2	# 2	# 3	# 4	0 4	0 5	1 7	2 9	↑												
	Second	125	250						# 2	0 3	0 3	1 5	1 6	3 8	4 10	7 14													
	Third	125	375						0 2	0 3	1 4	2 6	3 8	6 10	8 13	13 19													
	Fourth	125	500						0 3	1 4	2 5	3 7	5 10	8 13	12 17	19 25													
	Fifth	125	625						1 3	2 4	3 6	5 8	7 11	11 15	17 20	25 29													
	Sixth	125	750						1 3	3 5	4 6	7 9	10 12	14 17	21 23	31 33													
	Seventh	125	875						2 3	4 5	6 7	9 10	13 14	18 19	25 26	37 38													
P	First	200	200		•	↑	↓	# 2	# 2	# 3	# 4	0 4	0 5	1 7	2 9	↑													
	Second	200	400					# 2	0 3	0 3	1 5	1 6	3 8	4 10	7 14														
	Third	200	600					0 2	0 3	1 4	2 6	3 8	6 10	8 13	13 19														
	Fourth	200	800					0 3	1 4	2 5	3 7	5 10	8 13	12 17	19 25														
	Fifth	200	1000					1 3	2 4	3 6	5 8	7 11	11 15	17 20	25 29														
	Sixth	200	1200					1 3	3 5	4 6	7 9	10 12	14 17	21 23	31 33														
	Seventh	200	1400					2 3	4 5	6 7	9 10	13 14	18 19	25 26	37 38														
Q	First	315	315	•	↑	↓	# 2	# 2	# 3	# 4	0 4	0 5	1 7	2 9	↑														
	Second	315	630				# 2	0 3	0 3	1 5	1 6	3 8	4 10	7 14															
	Third	315	945				0 2	0 3	1 4	2 6	3 8	6 10	8 13	13 19															
	Fourth	315	1260				0 3	1 4	2 5	3 7	5 10	8 13	12 17	19 25															
	Fifth	315	1575				1 3	2 4	3 6	5 8	7 11	11 15	17 20	25 29															
	Sixth	315	1890				1 3	3 5	4 6	7 9	10 12	14 17	21 23	31 33															
	Seventh	315	2205				2 3	4 5	6 7	9 10	13 14	18 19	25 26	37 38															
R	First	500	500	↑	↑	# 2	# 2	# 3	# 4	0 4	0 5	1 7	2 9	↑															
	Second	500	1000			# 2	0 3	0 3	1 5	1 6	3 8	4 10	7 14																
	Third	500	1500			0 2	0 3	1 4	2 6	3 8	6 10	8 13	13 19																
	Fourth	500	2000			0 3	1 4	2 5	3 7	5 10	8 13	12 17	19 25																
	Fifth	500	2500			1 3	2 4	3 6	5 8	7 11	11 15	17 20	25 29																
	Sixth	500	3000			1 3	3 5	4 6	7 9	10 12	14 17	21 23	31 33																
	Seventh	500	3500			2 3	4 5	6 7	9 10	13 14	18 19	25 26	37 38																

↓ = Use first sampling plan below arrow. If sample size equals or exceeds lot or batch size, do 100 percent inspection.
↑ = Use first sampling plan above arrow (refer to preceding page, when necessary).
Ac = Acceptance number.
Re = Rejection number.
• = Use corresponding single sampling plan (or alternatively, use multiple plan below, where available).
= Acceptance not permitted at this sample size.

Table IV-B Multiple Sampling Plans for Tightened Inspection (Master Table).

Sample size code letter	Sample	Sample size	Cumulative sample size	Acceptable Quality Levels (tightened inspection)																									
				0.010	0.015	0.025	0.040	0.065	0.10	0.15	0.25	0.40	0.65	1.0	1.5	2.5	4.0	6.5	10	15	25	40	65	100	150	250	400	650	1000
				Ac Re	Ac Re	Ac Re	Ac Re	Ac Re	Ac Re	Ac Re	Ac Re	Ac Re	Ac Re	Ac Re	Ac Re	Ac Re	Ac Re	Ac Re	Ac Re	Ac Re	Ac Re	Ac Re	Ac Re	Ac Re	Ac Re	Ac Re	Ac Re	Ac Re	Ac Re
A B C				↓	↓	↓	↓	↓	↓	↓	↓	↓	↓	↓	↓	↓	↓ •	↓ •	↓	↓ ++	↓ ++ ++	• ++ ++	• ++ ++	• ++ ++	• ++ ++	• ++ ++	• ++ ++	• ++ ++	• ++ ↑
D	First	2	2	↓	↓	↓	↓	↓	↓	↓	↓	↓	↓	↓	↓	•	↓	↓	# 2	# 2	# 3	# 4	0 4	0 6	1 8	3 10	6 15	↑	↑
	Second	2	4																# 2	0 3	0 3	1 5	2 7	3 9	6 12	10 17	16 25		
	Third	2	6																0 2	0 3	1 4	2 6	4 9	7 12	11 17	17 24	26 36		
	Fourth	2	8																0 3	1 4	2 5	3 7	6 11	10 15	16 22	24 31	37 46		
	Fifth	2	10																1 3	2 4	3 6	5 8	9 12	14 17	22 25	32 37	49 55		
	Sixth	2	12																1 3	3 5	4 6	7 9	12 14	18 20	27 29	40 43	61 64		
	Seventh	2	14																2 3	4 5	6 7	9 10	14 15	21 22	32 33	48 49	72 73		
E	First	3	3	↓	↓	↓	↓	↓	↓	↓	↓	↓	↓	↓	•	↓	↓	# 2	# 2	# 3	# 4	0 4	0 6	1 8	3 10	6 15	↑	↑	↑
	Second	3	6															# 2	0 3	0 3	1 5	2 7	3 9	6 12	10 17	16 25			
	Third	3	9															0 2	0 3	1 4	2 6	4 9	7 12	11 17	17 24	26 36			
	Fourth	3	12															0 3	1 4	2 5	3 7	6 11	10 15	16 22	24 31	37 46			
	Fifth	3	15															1 3	2 4	3 6	5 8	9 12	14 17	22 25	32 37	49 55			
	Sixth	3	18															1 3	3 5	4 6	7 9	12 14	18 20	27 29	40 43	61 64			
	Seventh	3	21															2 3	4 5	6 7	9 10	14 15	21 22	32 33	48 49	72 73			
F	First	5	5	↓	↓	↓	↓	↓	↓	↓	↓	↓	↓	•	↓	↓	# 2	# 2	# 3	# 4	0 4	0 6	1 8	↑	↑	↑	↑	↑	↑
	Second	5	10														# 2	0 3	0 3	1 5	2 7	3 9	6 12						
	Third	5	15														0 2	0 3	1 4	2 6	4 9	7 12	11 17						
	Fourth	5	20														0 3	1 4	2 5	3 7	6 11	10 15	16 22						
	Fifth	5	25														1 3	2 4	3 6	5 8	9 12	14 17	22 25						
	Sixth	5	30														1 3	3 5	4 6	7 9	12 14	18 20	27 29						
	Seventh	5	35														2 3	4 5	6 7	9 10	14 15	21 22	32 33						
G	First	8	8	↓	↓	↓	↓	↓	↓	↓	↓	↓	•	↓	↓	# 2	# 2	# 3	# 4	0 4	0 6	1 8	↑	↑	↑	↑	↑	↑	↑
	Second	8	16													# 2	0 3	0 3	1 5	2 7	3 9	6 12							
	Third	8	24													0 2	0 3	1 4	2 6	4 9	7 12	11 17							
	Fourth	8	32													0 3	1 4	2 5	3 7	6 11	10 15	16 22							
	Fifth	8	40													1 3	2 4	3 6	5 8	9 12	14 17	22 25							
	Sixth	8	48													1 3	3 5	4 6	7 9	12 14	18 20	27 29							
	Seventh	8	56													2 3	4 5	6 7	9 10	14 15	21 22	32 33							
H	First	13	13	↓	↓	↓	↓	↓	↓	↓	↓	•	↓	↓	# 2	# 2	# 3	# 4	0 4	0 6	1 8	↑	↑	↑	↑	↑	↑	↑	↑
	Second	13	26												# 2	0 3	0 3	1 5	2 7	3 9	6 12								
	Third	13	39												0 2	0 3	1 4	2 6	4 9	7 12	11 17								
	Fourth	13	52												0 3	1 4	2 5	3 7	6 11	10 15	16 22								
	Fifth	13	65												1 3	2 4	3 6	5 8	9 12	14 17	22 25								
	Sixth	13	78												1 3	3 5	4 6	7 9	12 14	18 20	27 29								
	Seventh	13	91												2 3	4 5	6 7	9 10	14 15	21 22	32 33								
J	First	20	20	↓	↓	↓	↓	↓	↓	↓	•	↓	↓	# 2	# 2	# 3	# 4	0 4	0 6	1 8	↑	↑	↑	↑	↑	↑	↑	↑	↑
	Second	20	40											# 2	0 3	0 3	1 5	2 7	3 9	6 12									
	Third	20	60											0 2	0 3	1 4	2 6	4 9	7 12	11 17									
	Fourth	20	80											0 3	1 4	2 5	3 7	6 11	10 15	16 22									
	Fifth	20	100											1 3	2 4	3 6	5 8	9 12	14 17	22 25									
	Sixth	20	120											1 3	3 5	4 6	7 9	12 14	18 20	27 29									
	Seventh	20	140											2 3	4 5	6 7	9 10	14 15	21 22	32 33									

Table IV-B (continued).

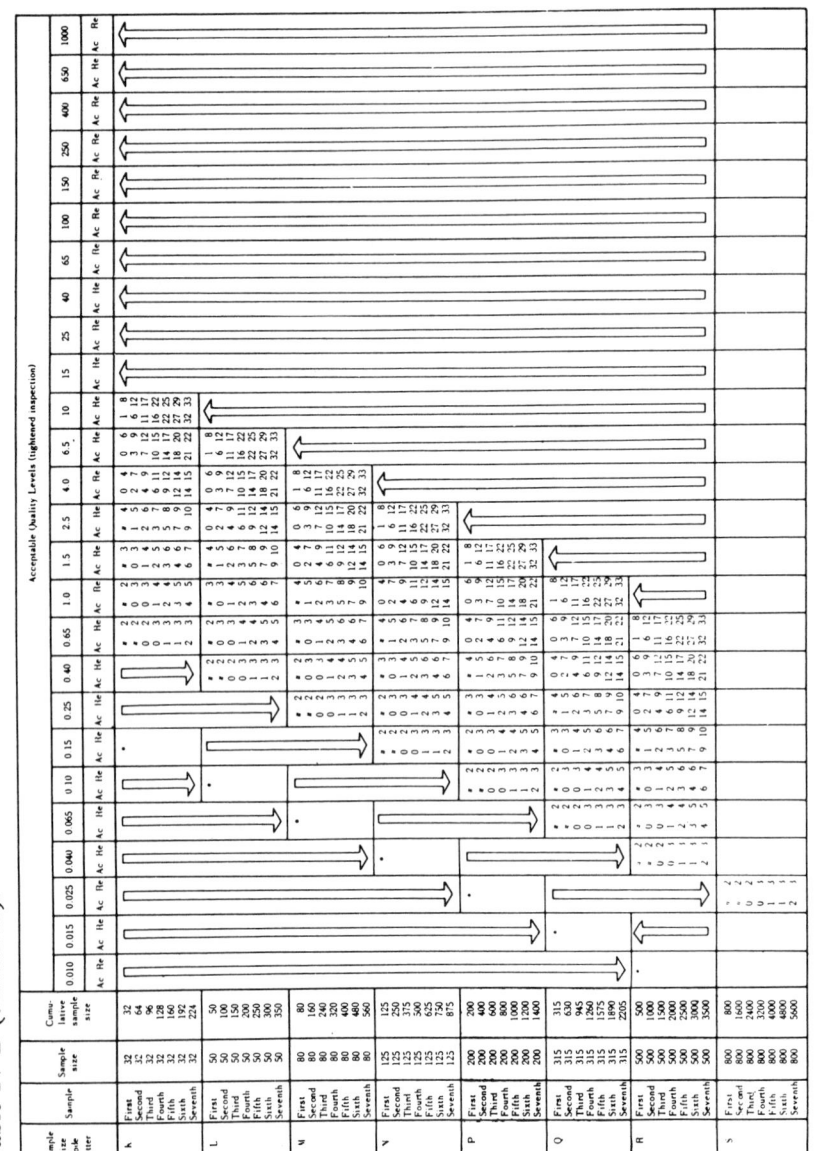

Sample size code letter	Sample	Sample size	Cumulative sample size	Acceptable Quality Levels (tightened inspection)																									
				0.010	0.015	0.025	0.040	0.065	0.10	0.15	0.25	0.40	0.65	1.0	1.5	2.5	4.0	6.5	10	15	25	40	65	100	150	250	400	650	1000
				Ac Re	Ac Re	Ac Re	Ac Re	Ac Re	Ac Re	Ac Re	Ac Re	Ac Re	Ac Re	Ac Re	Ac Re	Ac Re	Ac Re	Ac Re	Ac Re	Ac Re	Ac Re	Ac Re	Ac Re	Ac Re	Ac Re	Ac Re	Ac Re	Ac Re	Ac Re
K	First	32	32	↓	↓	↓	↓	↓	↓	•	↓	↓	* 2	* 2	* 3	* 4	0 4	0 6	1 8	↑	↑	↑	↑	↑	↑	↑	↑	↑	↑
	Second	32	64										* 2	0 3	0 3	1 5	2 7	3 9	6 12										
	Third	32	96										0 2	0 3	1 4	2 6	4 9	7 12	11 17										
	Fourth	32	128										0 3	1 4	2 5	3 7	6 11	10 15	16 22										
	Fifth	32	160										1 3	2 4	3 6	5 8	9 12	14 17	22 25										
	Sixth	32	192										1 3	3 5	4 6	7 9	12 14	18 20	27 29										
	Seventh	32	224										2 3	4 5	6 7	9 10	14 15	21 22	32 33										
L	First	50	50	↓	↓	↓	↓	↓	•	↓	↓	* 2	* 2	* 3	* 4	0 4	0 6	1 8	↑	↑	↑	↑	↑	↑	↑	↑	↑	↑	↑
	Second	50	100									* 2	0 3	0 3	1 5	2 7	3 9	6 12											
	Third	50	150									0 2	0 3	1 4	2 6	4 9	7 12	11 17											
	Fourth	50	200									0 3	1 4	2 5	3 7	6 11	10 15	16 22											
	Fifth	50	250									1 3	2 4	3 6	5 8	9 12	14 17	22 25											
	Sixth	50	300									1 3	3 5	4 6	7 9	12 14	18 20	27 29											
	Seventh	50	350									2 3	4 5	6 7	9 10	14 15	21 22	32 33											
M	First	80	80	↓	↓	↓	↓	•	↓	↓	* 2	* 2	* 3	* 4	0 4	0 6	1 8	↑	↑	↑	↑	↑	↑	↑	↑	↑	↑	↑	↑
	Second	80	160								* 2	0 3	0 3	1 5	2 7	3 9	6 12												
	Third	80	240								0 2	0 3	1 4	2 6	4 9	7 12	11 17												
	Fourth	80	320								0 3	1 4	2 5	3 7	6 11	10 15	16 22												
	Fifth	80	400								1 3	2 4	3 6	5 8	9 12	14 17	22 25												
	Sixth	80	480								1 3	3 5	4 6	7 9	12 14	18 20	27 29												
	Seventh	80	560								2 3	4 5	6 7	9 10	14 15	21 22	32 33												
N	First	125	125	↓	↓	↓	•	↓	↓	* 2	* 2	* 3	* 4	0 4	0 6	1 8	↑	↑	↑	↑	↑	↑	↑	↑	↑	↑	↑	↑	↑
	Second	125	250							* 2	0 3	0 3	1 5	2 7	3 9	6 12													
	Third	125	375							0 2	0 3	1 4	2 6	4 9	7 12	11 17													
	Fourth	125	500							0 3	1 4	2 5	3 7	6 11	10 15	16 22													
	Fifth	125	625							1 3	2 4	3 6	5 8	9 12	14 17	22 25													
	Sixth	125	750							1 3	3 5	4 6	7 9	12 14	18 20	27 29													
	Seventh	125	875							2 3	4 5	6 7	9 10	14 15	21 22	32 33													
P	First	200	200	↓	↓	•	↓	↓	* 2	* 2	* 3	* 4	0 4	0 6	1 8	↑	↑	↑	↑	↑	↑	↑	↑	↑	↑	↑	↑	↑	↑
	Second	200	400						* 2	0 3	0 3	1 5	2 7	3 9	6 12														
	Third	200	600						0 2	0 3	1 4	2 6	4 9	7 12	11 17														
	Fourth	200	800						0 3	1 4	2 5	3 7	6 11	10 15	16 22														
	Fifth	200	1000						1 3	2 4	3 6	5 8	9 12	14 17	22 25														
	Sixth	200	1200						1 3	3 5	4 6	7 9	12 14	18 20	27 29														
	Seventh	200	1400						2 3	4 5	6 7	9 10	14 15	21 22	32 33														
Q	First	315	315	↓	•	↓	↓	* 2	* 2	* 3	* 4	0 4	0 6	1 8	↑	↑	↑	↑	↑	↑	↑	↑	↑	↑	↑	↑	↑	↑	↑
	Second	315	630					* 2	0 3	0 3	1 5	2 7	3 9	6 12															
	Third	315	945					0 2	0 3	1 4	2 6	4 9	7 12	11 17															
	Fourth	315	1260					0 3	1 4	2 5	3 7	6 11	10 15	16 22															
	Fifth	315	1575					1 3	2 4	3 6	5 8	9 12	14 17	22 25															
	Sixth	315	1890					1 3	3 5	4 6	7 9	12 14	18 20	27 29															
	Seventh	315	2205					2 3	4 5	6 7	9 10	14 15	21 22	32 33															
R	First	500	500	•	↑	↓	* 2	* 2	* 3	* 4	0 4	0 6	1 8	↑	↑	↑	↑	↑	↑	↑	↑	↑	↑	↑	↑	↑	↑	↑	↑
	Second	500	1000				* 2	0 3	0 3	1 5	2 7	3 9	6 12																
	Third	500	1500				0 2	0 3	1 4	2 6	4 9	7 12	11 17																
	Fourth	500	2000				0 3	1 4	2 5	3 7	6 11	10 15	16 22																
	Fifth	500	2500				1 3	2 4	3 6	5 8	9 12	14 17	22 25																
	Sixth	500	3000				1 3	3 5	4 6	7 9	12 14	18 20	27 29																
	Seventh	500	3500				2 3	4 5	6 7	9 10	14 15	21 22	32 33																
S	First	800	800			* 2																							
	Second	800	1600			* 2																							
	Third	800	2400			0 2																							
	Fourth	800	3200			0 3																							
	Fifth	800	4000			1 3																							
	Sixth	800	4800			1 3																							
	Seventh	800	5600			2 3																							

⇩ = Use first sampling plan below arrow. If sample size equals or exceeds lot or batch size, do 100 percent inspection.

⇧ = Use first sampling plan above arrow (refer to preceding page, when necessary).

Ac = Acceptance number

Re = Rejection number

• = Use corresponding single sampling plan (or alternatively, use multiple sampling plan below, where available).

* = Acceptance not permitted at this sample size.

Table V-B Average Outgoing Quality Limit Factors for Tightened Inspection (Single Sampling).

Code letter	Sample size	Acceptable Quality Level																									
		0.010	0.015	0.025	0.040	0.065	0.10	0.15	0.25	0.40	0.65	1.0	1.5	2.5	4.0	6.5	10	15	25	40	65	100	150	250	400	650	1000
A	2																			42	69	97	160	260	400	620	970
B	3															12			28	46	65	110	170	270	410	650	1100
C	5														7.4			17	27	39	63	100	160	250	390	610	
D	8													4.6			11	17	24	40	64	99	160	240	380		
E	13												2.8			6.5	11	15	24	40	61	95	150	240			
F	20											1.8			4.2	6.9	9.7	16	26	40	62						
G	32										1.2			2.6	4.3	6.1	9.9	16	25	39							
H	50									0.74			1.7	2.7	3.9	6.3	10	16	25								
J	80								0.46			1.1	1.7	2.4	4.0	6.4	9.9	16									
K	125							0.29			0.67	1.1	1.6	2.5	4.1	6.4	9.9										
L	200						0.18			0.42	0.69	0.97	1.6	2.6	4.0	6.2											
M	315					0.12			0.27	0.44	0.62	1.0	1.6	2.5	3.9												
N	500				0.074			0.17	0.27	0.39	0.63	1.0	1.6	2.5													
P	800			0.046			0.11	0.17	0.24	0.40	0.64	0.99	1.6														
Q	1250		0.029			0.067	0.11	0.16	0.25	0.41	0.64	0.99															
R	2000	0.018			0.042	0.069	0.097	0.16	0.26	0.40	0.62																
S	3150			0.027																							

Note: For the exact AOQL, the above values must be multiplied by $\left(1 - \frac{\text{Sample size}}{\text{Lot or Batch size}}\right)$ (see 11.4)

Table VI-A Limiting Quality (in Percent Defective) for Which P_a = 10 Percent (for Normal Inspection, Single Sampling).

Code letter	Sample size	Acceptable Quality Level															
		0.010	0.015	0.025	0.040	0.065	0.10	0.15	0.25	0.40	0.65	1.0	1.5	2.5	4.0	6.5	10
A	2															68	
B	3														54		
C	5													37			58
D	8												25			41	54
E	13											16			27	36	44
F	20										11			18	25	30	42
G	32									6.9			12	16	20	27	34
H	50								4.5			7.6	10	13	18	22	29
J	80							2.8			4.8	6.5	8.2	11	14	19	24
K	125						1.8			3.1	4.3	5.4	7.4	9.4	12	16	23
L	200					1.2			2.0	2.7	3.3	4.6	5.9	7.7	10	14	
M	315				0.73			1.2	1.7	2.1	2.9	3.7	4.9	6.4	9.0		
N	500			0.46			0.78	1.1	1.3	1.9	2.4	3.1	4.0	5.6			
P	800		0.29			0.49	0.67	0.84	1.2	1.5	1.9	2.5	3.5				
Q	1250	0.18			0.31	0.43	0.53	0.74	0.94	1.2	1.6	2.3					
R	2000			0.20	0.27	0.33	0.46	0.59	0.77	1.0	1.4						

Table VI-B Limiting Quality (in Defects per Hundred Units) for Which P_a = 10 Percent (for Normal Inspection, Single Sampling).

Code letter	Sample size	Acceptable Quality Level																									
		0.010	0.015	0.025	0.040	0.065	0.10	0.15	0.25	0.40	0.65	1.0	1.5	2.5	4.0	6.5	10	15	25	40	65	100	150	250	400	650	1000
A	2															120			200	270	330	460	590	770	1000	1400	1900
B	3														77			130	180	220	310	390	510	670	940	1300	1800
C	5													46			78	110	130	190	240	310	400	560	770	1100	
D	8												29			49	67	84	120	150	190	250	350	480	670		
E	13											18			30	41	51	71	91	120	160	220	300	410			
F	20										12			20	27	33	46	59	77	100	140						
G	32									7.2			12	17	21	29	37	48	63	88							
H	50								4.6			7.8	11	13	19	24	31	40	56								
J	80							2.9			4.9	6.7	8.4	12	15	19	25	35									
K	125						1.8			3.1	4.3	5.4	7.4	9.4	12	16	23										
L	200					1.2			2.0	2.7	3.3	4.6	5.9	7.7	10	14											
M	315				0.73			1.2	1.7	2.1	2.9	3.7	4.9	6.4	9.0												
N	500			0.46			0.78	1.1	1.3	1.9	2.4	3.1	4.0	5.6													
P	800		0.29			0.49	0.67	0.84	1.2	1.5	1.9	2.5	3.5														
Q	1250	0.18			0.31	0.43	0.53	0.74	0.94	1.2	1.6	2.3															
R	2000			0.20	0.27	0.33	0.46	0.59	0.77	1.0	1.4																

Table VIII Limit Numbers for Reduced Inspection.

Number of sample units from last 10 lots or batches	Acceptable Quality Level																									
	0.010	0.015	0.025	0.040	0.065	0.10	0.15	0.25	0.40	0.65	1.0	1.5	2.5	4.0	6.5	10	15	25	40	65	100	150	250	400	650	1000
20 - 29	•	•	•	•	•	•	•	•	•	•	•	•	•	•	•	0	0	2	4	8	14	22	40	68	115	181
30 - 49	•	•	•	•	•	•	•	•	•	•	•	•	•	•	0	0	1	3	7	13	22	36	63	105	178	277
50 - 79	•	•	•	•	•	•	•	•	•	•	•	•	•	0	0	2	3	7	14	25	40	63	110	181	301	
80 - 129	•	•	•	•	•	•	•	•	•	•	•	•	0	0	2	4	7	14	24	42	68	105	181	297		
130 - 199	•	•	•	•	•	•	•	•	•	•	•	0	0	2	4	7	13	25	42	72	115	177	301	490		
200 - 319	•	•	•	•	•	•	•	•	•	•	0	0	2	4	8	14	22	40	68	115	181	277	471			
320 - 499	•	•	•	•	•	•	•	•	•	0	0	1	4	8	14	24	39	68	113	189						
50C - 799	•	•	•	•	•	•	•	•	0	0	2	3	7	14	25	40	63	110	181							
800 - 1249	•	•	•	•	•	•	•	0	0	2	4	7	14	24	42	68	105	181								
1250 - 1999	•	•	•	•	•	•	0	0	2	4	7	13	24	40	69	110	169									
2000 - 3149	•	•	•	•	•	0	0	2	4	8	14	22	40	68	115	181										
3150 - 4999	•	•	•	•	0	0	1	4	8	14	24	38	67	111	186											
5000 - 7999	•	•	•	0	0	2	3	7	14	25	40	63	110	181												
8000 - 12499	•	•	0	0	2	4	7	14	24	42	68	105	181													
12500 - 19999	•	0	0	2	4	7	13	24	40	69	110	169														
20000 - 31499	0	0	2	4	8	14	22	40	68	115	181															
31500 - 49999	0	1	4	8	14	24	38	67	111	186																
50000 & Over	2	3	7	14	25	40	63	110	181	301																

Denotes that the number of sample units from the last ten lots or batches is not sufficient for reduced inspection for this AQL. In this instance more than ten lots or batches may be used for the calculation, provided that the lots or batches used are the most recent ones in sequence, that they have all been on normal inspection, and that none has been rejected while on original inspection.

Table IX Average Sample Size Curves for Double and Multiple Sampling (Normal and Tightened Inspection).

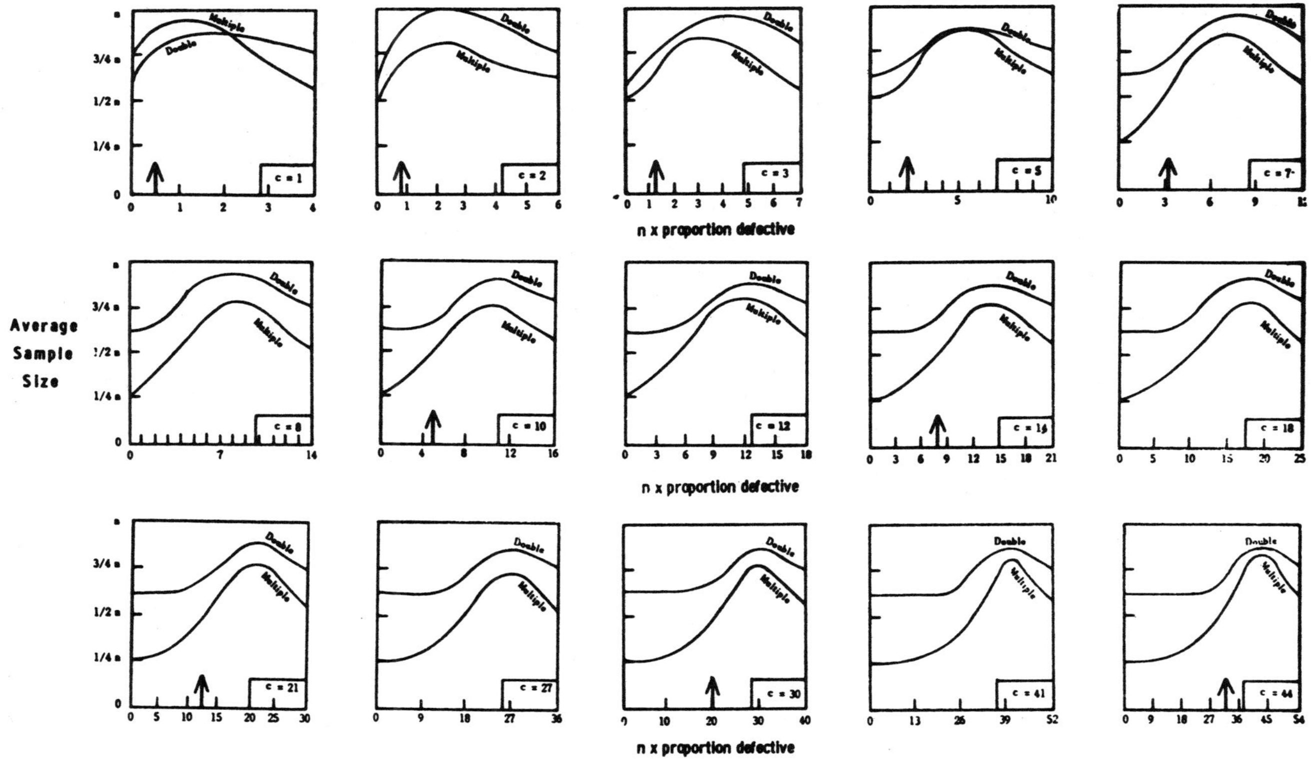

n = Equivalent single sample size
c = Single sample acceptance number
↑ = AQL for normal inspection

Table X-J Tables for Sample Size Code Letter: J.

CHART J - OPERATING CHARACTERISTIC CURVES FOR SINGLE SAMPLING PLANS

(Curves for double and multiple sampling are matched as closely as practicable)

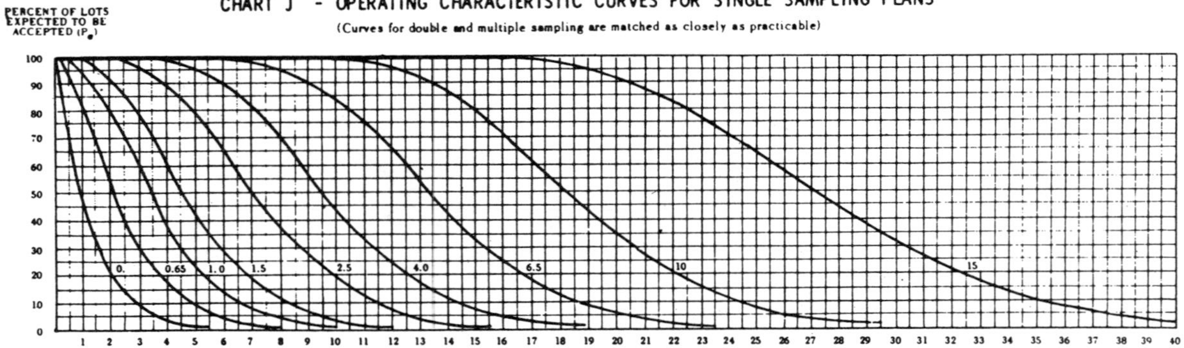

QUALITY OF SUBMITTED LOTS (p, in percent defective for AQL's ≤10; in defects per hundred units for AQL's >10)

Note: Figures on curves are Acceptable Quality Levels (AQL's) for normal inspection.

Table X-J-I Tabulated Values for Operating Characteristic Curves for Single Sampling Plans.

P_a	Acceptable Quality Levels (normal inspection)																					
	0.15	0.65	1.0	1.5	2.5	4.0	X	6.5	X	10	0.15	0.65	1.0	1.5	2.5	4.0	X	6.5	X	10	X	15
	p (in percent defective)										p (in defects per hundred units)											
99.0	0.013	0.188	0.550	1.05	2.30	3.72	4.50	6.13	7.88	9.75	0.013	0.186	0.545	1.03	2.23	3.63	4.38	5.96	7.62	9.35	12.9	15.7
95.0	0.064	0.444	1.03	1.73	3.32	5.06	5.98	7.91	9.89	11.9	0.064	0.444	1.02	1.71	3.27	4.98	5.87	7.71	9.61	11.6	15.6	18.6
90.0	0.132	0.666	1.38	2.20	3.98	5.91	6.91	8.95	11.0	13.2	0.131	0.665	1.38	2.18	3.94	5.82	6.79	8.78	10.8	12.9	17.1	20.3
75.0	0.359	1.202	2.16	3.18	5.30	7.50	8.62	10.9	13.2	15.5	0.360	1.20	2.16	3.17	5.27	7.45	8.55	10.8	13.0	15.3	19.9	23.4
50.0	0.863	2.09	3.33	4.57	7.06	9.55	10.8	13.3	15.8	18.3	0.866	2.10	3.34	4.59	7.09	9.59	10.8	13.3	15.8	18.3	23.3	27.1
25.0	1.72	3.33	4.84	6.31	9.14	11.9	13.3	16.0	18.6	21.3	1.73	3.37	4.90	6.39	9.28	12.1	13.5	16.3	19.0	21.8	27.2	31.2
10.0	2.84	4.78	6.52	8.16	11.3	14.2	15.7	18.6	21.4	24.2	2.88	4.86	6.65	8.35	11.6	14.7	16.2	19.3	22.2	25.2	30.9	35.2
5.0	3.68	5.80	7.66	9.39	12.7	15.8	17.3	20.3	23.2	26.0	3.75	5.93	7.87	9.69	13.1	16.4	18.0	21.2	24.3	27.4	33.4	37.8
1.0	5.59	8.00	10.1	12.0	15.6	18.9	20.5	23.6	26.5	29.5	5.76	8.30	10.5	12.6	16.4	20.0	21.8	25.2	28.5	31.8	38.2	42.9
	0.25	1.0	1.5	2.5	4.0	X	6.5	X	10	X	0.25	1.0	1.5	2.5	4.0	X	6.5	X	10	X	15	X
	Acceptable Quality Levels (tightened inspection)																					

Note: All values given in above table based on Poisson distribution as an approximation to the Binomial.

Table X-J-2 Sampling Plans for Sample Size Code Letter: J.

Type of sampling plan	Cumulative sample size	Acceptable Quality Levels (normal inspection)																																		Cumulative sample size
		Less than 0.15		0.15		0.25		X		0.40		0.65		1.0		1.5		2.5		4.0		X		6.5		X		10		X		15		Higher than 15		
		Ac	Re	Ac	Re	Ac	Re	Ac	Re	Ac	Re	Ac	Re	Ac	Re	Ac	Re	Ac	Re	Ac	Re	Ac	Re	Ac	Re	Ac	Re	Ac	Re	Ac	Re	Ac	Re	Ac	Re	
Single	80	▽		0	1							1	2	2	3	3	4	5	6	7	8	8	9	10	11	12	13	14	15	18	19	21	22	△		80
Double	50	▽		•		Use Letter H		Use Letter L		Use Letter K		0	2	0	3	1	4	2	5	3	7	3	7	5	9	6	10	7	11	9	14	11	16	△		50
	100											1	2	3	4	4	5	6	7	8	9	11	12	12	13	15	16	18	19	23	24	26	27			100
Multiple	20	▽		•								#	2	#	2	#	3	#	4	0	4	0	4	0	5	0	6	1	7	1	8	2	9	△		20
	40											#	2	0	3	0	3	1	5	1	6	2	7	3	8	3	9	4	10	6	12	7	14			40
	60											0	2	0	3	1	4	2	6	3	8	4	9	6	10	7	12	8	13	11	17	13	19			60
	80											0	3	1	4	2	5	3	7	5	10	6	11	8	13	10	15	12	17	16	22	19	25			80
	100											1	3	2	4	3	6	5	8	7	11	9	12	11	15	14	17	17	20	22	25	25	29			100
	120											1	3	3	5	4	6	7	9	10	12	12	14	14	17	18	20	21	23	27	29	31	33			120
	140											2	3	4	5	6	7	9	10	13	14	14	15	18	19	21	22	25	26	32	33	37	38			140
		Less than 0.25		0.25		X		0.40		0.65		1.0		1.5		2.5		4.0		X		6.5		X		10		X		15		X		Higher than 15		
		Acceptable Quality Levels (tightened inspection)																																		

△ = Use next preceding sample size code letter for which acceptance and rejection numbers are available.
▽ = Use next subsequent sample size code letter for which acceptance and rejection numbers are available.
Ac = Acceptance number
Re = Rejection number
• = Use single sampling plan above (or alternatively use letter M)
= Acceptance not permitted at this sample size.

Table 1 Cumulative Normal Distribution--Values of *P*.

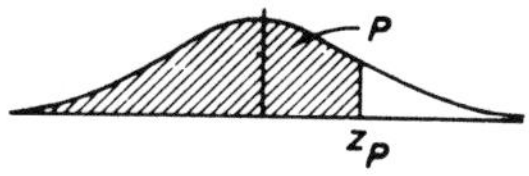

Values of *P* corresponding to z_P for the normal curve.

z is the standard normal variable. The value of *P* for $-z_P$ equals one minus the value of *P* for $+z_P$, e.g., the *P* for -1.62 equals $1 - .9474 = .0526$.

z_P	.00	.01	.02	.03	.04	.05	.06	.07	.08	.09
.0	.5000	.5040	.5080	.5120	.5160	.5199	.5239	.5279	.5319	.5359
.1	.5398	.5438	.5478	.5517	.5557	.5596	.5636	.5675	.5714	.5753
.2	.5793	.5832	.5871	.5910	.5948	.5987	.6026	.6064	.6103	.6141
.3	.6179	.6217	.6255	.6293	.6331	.6368	.6406	.6443	.6480	.6517
.4	.6554	.6591	.6628	.6664	.6700	.6736	.6772	.6808	.6844	.6879
.5	.6915	.6950	.6985	.7019	.7054	.7088	.7123	.7157	.7190	.7224
.6	.7257	.7291	.7324	.7357	.7389	.7422	.7454	.7486	.7517	.7549
.7	.7580	.7611	.7642	.7673	.7704	.7734	.7764	.7794	.7823	.7852
.8	.7881	.7910	.7939	.7967	.7995	.8023	.8051	.8078	.8106	.8133
.9	.8159	.8186	.8212	.8238	.8264	.8289	.8315	.8340	.8365	.8389
1.0	.8413	.8438	.8461	.8485	.8508	.8531	.8554	.8577	.8599	.8621
1.1	.8643	.8665	.8686	.8708	.8729	.8749	.8770	.8790	.8810	.8830
1.2	.8849	.8869	.8888	.8907	.8925	.8944	.8962	.8980	.8997	.9015
1.3	.9032	.9049	.9066	.9082	.9099	.9115	.9131	.9147	.9162	.9177
1.4	.9192	.9207	.9222	.9236	.9251	.9265	.9279	.9292	.9306	.9319
1.5	.9332	.9345	.9357	.9370	.9382	.9394	.9406	.9418	.9429	.9441
1.6	.9452	.9463	.9474	.9484	.9495	.9505	.9515	.9525	.9535	.9545
1.7	.9554	.9564	.9573	.9582	.9591	.9599	.9608	.9616	.9625	.9633
1.8	.9641	.9649	.9656	.9664	.9671	.9678	.9686	.9693	.9699	.9706
1.9	.9713	.9719	.9726	.9732	.9738	.9744	.9750	.9756	.9761	.9767
2.0	.9772	.9778	.9783	.9788	.9793	.9798	.9803	.9808	.9812	.9817
2.1	.9821	.9826	.9830	.9834	.9838	.9842	.9846	.9850	.9854	.9857
2.2	.9861	.9864	.9868	.9871	.9875	.9878	.9881	.9884	.9887	.9890
2.3	.9893	.9896	.9898	.9901	.9904	.9906	.9909	.9911	.9913	.9916
2.4	.9918	.9920	.9922	.9925	.9927	.9929	.9931	.9932	.9934	.9936
2.5	.9938	.9940	.9941	.9943	.9945	.9946	.9948	.9949	.9951	.9952
2.6	.9953	.9955	.9956	.9957	.9959	.9960	.9961	.9962	.9963	.9964
2.7	.9965	.9966	.9967	.9968	.9969	.9970	.9971	.9972	.9973	.9974
2.8	.9974	.9975	.9976	.9977	.9977	.9978	.9979	.9979	.9980	.9981
2.9	.9981	.9982	.9982	.9983	.9984	.9984	.9985	.9985	.9986	.9986
3.0	.9987	.9987	.9987	.9988	.9988	.9989	.9989	.9989	.9990	.9990
3.1	.9990	.9991	.9991	.9991	.9992	.9992	.9992	.9992	.9993	.9993
3.2	.9993	.9993	.9994	.9994	.9994	.9994	.9994	.9995	.9995	.9995
3.3	.9995	.9995	.9995	.9996	.9996	.9996	.9996	.9996	.9996	.9997
3.4	.9997	.9997	.9997	.9997	.9997	.9997	.9997	.9997	.9997	.9998

Table 2 Poisson Distribution. Probabilities of c or less, given c ′, appears in the body of the table multiplied by 1000.

c' or np' \ c	0	1	2	3	4	5	6	7	8	9
0.02	980	1,000								
0.04	961	999	1,000							
0.06	942	998	1,000							
0.08	923	997	1,000							
0.10	905	995	1,000							
0.15	861	990	999	1,000						
0.20	819	982	999	1,000						
0.25	779	974	998	1,000						
0.30	741	963	996	1,000						
0.35	705	951	994	1,000						
0.40	670	938	992	999	1,000					
0.45	638	925	989	999	1,000					
0.50	607	910	986	998	1,000					
0.55	577	894	982	998	1,000					
0.60	549	878	977	997	1,000					
0.65	522	861	972	996	999	1,000				
0.70	497	844	966	994	999	1,000				
0.75	472	827	959	993	999	1,000				
0.80	449	809	953	991	999	1,000				
0.85	427	791	945	989	998	1,000				
0.90	407	772	937	987	998	1,000				
0.95	387	754	929	984	997	1,000				
1.00	368	736	920	981	996	999	1,000			
1.1	333	699	900	974	995	999	1,000			
1.2	301	663	879	966	992	998	1,000			
1.3	273	627	857	957	989	998	1,000			
1.4	247	592	833	946	986	997	999	1,000		
1.5	223	558	809	934	981	996	999	1,000		
1.6	202	525	783	921	976	994	999	1,000		
1.7	183	493	757	907	970	992	998	1,000		
1.8	165	463	731	891	964	990	997	999	1,000	
1.9	150	434	704	875	956	987	997	999	1,000	
2.0	135	406	677	857	947	983	995	999	1,000	

c' or np' \ c	0	1	2	3	4	5	6	7	8	9
2.2	111	355	623	819	928	975	993	998	1,000	
2.4	091	308	570	779	904	964	988	997	999	1,000
2.6	074	267	518	736	877	951	983	995	999	1,000
2.8	061	231	469	692	848	935	976	992	998	999
3.0	050	199	423	647	815	916	966	988	996	999
3.2	041	171	380	603	781	895	955	983	994	998
3.4	033	147	340	558	744	871	942	977	992	997
3.6	027	126	303	515	706	844	927	969	988	996
3.8	022	107	269	473	668	816	909	960	984	994
4.0	018	092	238	433	629	785	889	949	979	992
4.2	015	078	210	395	590	753	867	936	972	989
4.4	012	066	185	359	551	720	844	921	964	985
4.6	010	056	163	326	513	686	818	905	955	980
4.8	008	048	143	294	476	651	791	887	944	975
5.0	007	040	125	265	440	616	762	867	932	968
5.2	006	034	109	238	406	581	732	845	918	960
5.4	005	029	095	213	373	546	702	822	903	951
5.6	004	024	082	191	342	512	670	797	886	941
5.8	003	021	072	170	313	478	638	771	867	929
6.0	002	017	062	151	285	446	606	744	847	916

	10	11	12	13	14	15	16
2.8	1,000						
3.0	1,000						
3.2	1,000						
3.4	999	1,000					
3.6	999	1,000					
3.8	998	999	1,000				
4.0	997	999	1,000				
4.2	996	999	1,000				
4.4	994	998	999	1,000			
4.6	992	997	999	1,000			
4.8	990	996	999	1,000			
5.0	986	995	998	999	1,000		
5.2	982	993	997	999	1,000		
5.4	977	990	996	999	1,000		
5.6	972	988	995	998	999	1,000	
5.8	965	984	993	997	999	1,000	
6.0	957	980	991	996	999	999	1,000

Table 2 (continued).

c' or np' \ c	0	1	2	3	4	5	6	7	8	9
6.2	002	015	054	134	259	414	574	716	826	902
6.4	002	012	046	119	235	384	542	687	803	886
6.6	001	010	040	105	213	355	511	658	780	869
6.8	001	009	034	093	192	327	480	628	755	850
7.0	001	007	030	082	173	301	450	599	729	830
7.2	001	006	025	072	156	276	420	569	703	810
7.4	001	005	022	063	140	253	392	539	676	788
7.6	001	004	019	055	125	231	365	510	648	765
7.8	000	004	016	048	112	210	338	481	620	741
8.0	000	003	014	042	100	191	313	453	593	717
8.5	000	002	009	030	074	150	256	386	523	653
9.0	000	001	006	021	055	116	207	324	456	587
9.5	000	001	004	015	040	089	165	269	392	522
10.0	000	000	003	010	029	067	130	220	333	458

	10	11	12	13	14	15	16	17	18	19
6.2	949	975	989	995	998	999	1,000			
6.4	939	969	986	994	997	999	1,000			
6.6	927	963	982	992	997	999	999	1,000		
6.8	915	955	978	990	996	998	999	1,000		
7.0	901	947	973	987	994	998	999	1,000		
7.2	887	937	967	984	993	997	999	999	1,000	
7.4	871	926	961	980	991	996	998	999	1,000	
7.6	854	915	954	976	989	995	998	999	1,000	
7.8	835	902	945	971	986	993	997	999	1,000	
8.0	816	888	936	966	983	992	996	998	999	1,000
8.5	763	849	909	949	973	986	993	997	999	999
9.0	706	803	876	926	959	978	989	995	998	999
9.5	645	752	836	898	940	967	982	991	996	998
10.0	583	697	792	864	917	951	973	986	993	997

	20	21	22
8.5	1,000		
9.0	1,000		
9.5	999	1,000	
10.0	998	999	1,000

c' or np' \ c	0	1	2	3	4	5	6	7	8	9
10.5	000	000	002	007	021	050	102	179	279	397
11.0	000	000	001	005	015	038	079	143	232	341
11.5	000	000	001	003	011	028	060	114	191	289
12.0	000	000	001	002	008	020	046	090	155	242
12.5	000	000	000	002	005	015	035	070	125	201
13.0	000	000	000	001	004	011	026	054	100	166
13.5	000	000	000	001	003	008	019	041	079	135
14.0	000	000	000	000	002	006	014	032	062	109
14.5	000	000	000	000	001	004	010	024	048	088
15.0	000	000	000	000	001	003	008	018	037	070
	10	11	12	13	14	15	16	17	18	19
10.5	521	639	742	825	888	932	960	978	988	994
11.0	460	579	689	781	854	907	944	968	982	991
11.5	402	520	633	733	815	878	924	954	974	986
12.0	347	462	576	682	772	844	899	937	963	979
12.5	297	406	519	628	725	806	869	916	948	969
13.0	252	353	463	573	675	764	835	890	930	957
13.5	211	304	409	518	623	718	798	861	908	942
14.0	176	260	358	464	570	669	756	827	883	923
14.5	145	220	311	413	518	619	711	790	853	901
15.0	118	185	268	363	466	568	664	749	819	875
	20	21	22	23	24	25	26	27	28	29
10.5	997	999	999	1,000						
11.0	995	998	999	1,000						
11.5	992	996	998	999	1,000					
12.0	988	994	997	999	999	1,000				
12.5	983	991	995	998	999	999	1,000			
13.0	975	986	992	996	998	999	1,000			
13.5	965	980	989	994	997	998	999	1,000		
14.0	952	971	983	991	995	997	999	999	1,000	
14.5	936	960	976	986	992	996	998	999	999	1,000
15.0	917	947	967	981	989	994	997	998	999	1,000

Table 2 (continued).

c' or np' \ c	4	5	6	7	8	9	10	11	12	13
16	000	001	004	010	022	043	077	127	193	275
17	000	001	002	005	013	026	049	085	135	201
18	000	000	001	003	007	015	030	055	092	143
19	000	000	001	002	004	009	018	035	061	098
20	000	000	000	001	002	005	011	021	039	066
21	000	000	000	000	001	003	006	013	025	043
22	000	000	000	000	001	002	004	008	015	028
23	000	000	000	000	000	001	002	004	009	017
24	000	000	000	000	000	000	001	003	005	011
25	000	000	000	000	000	000	001	001	003	006
	14	15	16	17	18	19	20	21	22	23
16	368	467	566	659	742	812	868	911	942	963
17	281	371	468	564	655	736	805	861	905	937
18	208	287	375	469	562	651	731	799	855	899
19	150	215	292	378	469	561	647	725	793	849
20	105	157	221	297	381	470	559	644	721	787
21	072	111	163	227	302	384	471	558	640	716
22	048	077	117	169	232	306	387	472	556	637
23	031	052	082	123	175	238	310	389	472	555
24	020	034	056	087	128	180	243	314	392	473
25	012	022	038	060	092	134	185	247	318	394
	24	25	26	27	28	29	30	31	32	33
16	978	987	993	996	998	999	999	1,000		
17	959	975	985	991	995	997	999	999	1,000	
18	932	955	972	983	990	994	997	998	999	1,000
19	893	927	951	969	980	988	993	996	998	999
20	843	888	922	948	966	978	987	992	995	997
21	782	838	883	917	944	963	976	985	991	994
22	712	777	832	877	913	940	959	973	983	989
23	635	708	772	827	873	908	936	956	971	981
24	554	632	704	768	823	868	904	932	953	969
25	473	553	629	700	763	818	863	900	929	950
	34	35	36	37	38	39	40	41	42	43
19	999	1,000								
20	999	999	1,000							
21	997	998	999	999	1,000					
22	994	996	998	999	999	1,000				
23	988	993	996	997	999	999	1,000			
24	979	987	992	995	997	998	999	999	1,000	
25	966	978	985	991	994	997	998	999	999	1,000

Table 3 Percentiles of the *t* Distribution.

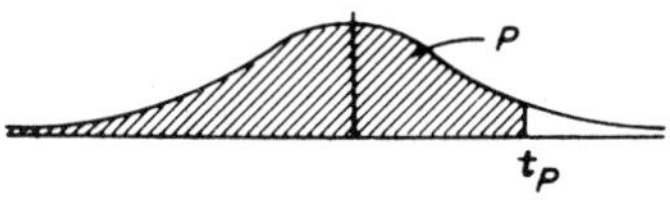

df	$t_{.60}$	$t_{.70}$	$t_{.80}$	$t_{.90}$	$t_{.95}$	$t_{.975}$	$t_{.99}$	$t_{.995}$
1	.325	.727	1.376	3.078	6.314	12.706	31.821	63.657
2	.289	.617	1.061	1.886	2.920	4.303	6.965	9.925
3	.277	.584	.978	1.638	2.353	3.182	4.541	5.841
4	.271	.569	.941	1.533	2.132	2.776	3.747	4.604
5	.267	.559	.920	1.476	2.015	2.571	3.365	4.032
6	.265	.553	.906	1.440	1.943	2.447	3.143	3.707
7	.263	.549	.896	1.415	1.895	2.365	2.998	3.499
8	.262	.546	.889	1.397	1.860	2.306	2.896	3.355
9	.261	.543	.883	1.383	1.833	2.262	2.821	3.250
10	.260	.542	.879	1.372	1.812	2.228	2.764	3.169
11	.260	.540	.876	1.363	1.796	2.201	2.718	3.106
12	.259	.539	.873	1.356	1.782	2.179	2.681	3.055
13	.259	.538	.870	1.350	1.771	2.160	2.650	3.012
14	.258	.537	.868	1.345	1.761	2.145	2.624	2.977
15	.258	.536	.866	1.341	1.753	2.131	2.602	2.947
16	.258	.535	.865	1.337	1.746	2.120	2.583	2.921
17	.257	.534	.863	1.333	1.740	2.110	2.567	2.898
18	.257	.534	.862	1.330	1.734	2.101	2.552	2.878
19	.257	.533	.861	1.328	1.729	2.093	2.539	2.861
20	.257	.533	.860	1.325	1.725	2.086	2.528	2.845
21	.257	.532	.859	1.323	1.721	2.080	2.518	2.831
22	.256	.532	.858	1.321	1.717	2.074	2.508	2.819
23	.256	.532	.858	1.319	1.714	2.069	2.500	2.807
24	.256	.531	.857	1.318	1.711	2.064	2.492	2.797
25	.256	.531	.856	1.316	1.708	2.060	2.485	2.787
26	.256	.531	.856	1.315	1.706	2.056	2.479	2.779
27	.256	.531	.855	1.314	1.703	2.052	2.473	2.771
28	.256	.530	.855	1.313	1.701	2.048	2.467	2.763
29	.256	.530	.854	1.311	1.699	2.045	2.462	2.756
30	.256	.530	.854	1.310	1.697	2.042	2.457	2.750
40	.255	.529	.851	1.303	1.684	2.021	2.423	2.704
60	.254	.527	.848	1.296	1.671	2.000	2.390	2.660
120	.254	.526	.845	1.289	1.658	1.980	2.358	2.617
∞	.253	.524	.842	1.282	1.645	1.960	2.326	2.576

Table 4 Percentiles of the F Distribution.

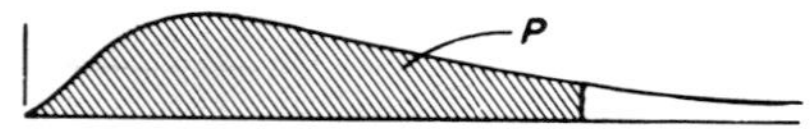

n_1 = degrees of freedom for numerator $\qquad F_{.90}(n_1, n_2) \qquad F_P$

n_2 = degrees of freedom for denominator

n_2 \ n_1	1	2	3	4	5	6	7	8	9	10	12	15	20	24	30	40	60	120	∞
1	39.86	49.50	53.59	55.83	57.24	58.20	58.91	59.44	59.86	60.19	60.71	61.22	61.74	62.00	62.26	62.53	62.79	63.06	63.33
2	8.53	9.00	9.16	9.24	9.29	9.33	9.35	9.37	9.38	9.39	9.41	9.42	9.44	9.45	9.46	9.47	9.47	9.48	9.49
3	5.54	5.46	5.39	5.34	5.31	5.28	5.27	5.25	5.24	5.23	5.22	5.20	5.18	5.18	5.17	5.16	5.15	5.14	5.13
4	4.54	4.32	4.19	4.11	4.05	4.01	3.98	3.95	3.94	3.92	3.90	3.87	3.84	3.83	3.82	3.80	3.79	3.78	3.76
5	4.06	3.78	3.62	3.52	3.45	3.40	3.37	3.34	3.32	3.30	3.27	3.24	3.21	3.19	3.17	3.16	3.14	3.12	3.10
6	3.78	3.46	3.29	3.18	3.11	3.05	3.01	2.98	2.96	2.94	2.90	2.87	2.84	2.82	2.80	2.78	2.76	2.74	2.72
7	3.59	3.26	3.07	2.96	2.88	2.83	2.78	2.75	2.72	2.70	2.67	2.63	2.59	2.58	2.56	2.54	2.51	2.49	2.47
8	3.46	3.11	2.92	2.81	2.73	2.67	2.62	2.59	2.56	2.50	2.50	2.46	2.42	2.40	2.38	2.36	2.34	2.32	2.29
9	3.36	3.01	2.81	2.69	2.61	2.55	2.51	2.47	2.44	2.42	2.38	2.34	2.30	2.28	2.25	2.23	2.21	2.18	2.16
10	3.29	2.92	2.73	2.61	2.52	2.46	2.41	2.38	2.35	2.32	2.28	2.24	2.20	2.18	2.16	2.13	2.11	2.08	2.06
11	3.23	2.86	2.66	2.54	2.45	2.39	2.34	2.30	2.27	2.25	2.21	2.17	2.12	2.10	2.08	2.05	2.03	2.00	1.97
12	3.18	2.81	2.61	2.48	2.39	2.33	2.28	2.24	2.21	2.19	2.15	2.10	2.06	2.04	2.01	1.99	1.96	1.93	1.90
13	3.14	2.76	2.56	2.43	2.35	2.28	2.23	2.20	2.16	2.14	2.10	2.05	2.01	1.98	1.96	1.93	1.90	1.88	1.85
14	3.10	2.73	2.52	2.39	2.31	2.24	2.19	2.15	2.12	2.10	2.05	2.01	1.96	1.94	1.9[illegible]	1.89	1.86	1.83	1.80
15	3.07	2.70	2.49	2.36	2.27	2.21	2.16	2.12	2.09	2.06	2.02	1.97	1.92	1.90	1.87	1.85	1.82	1.79	1.76
16	3.05	2.67	2.46	2.33	2.24	2.18	2.13	2.09	2.06	2.03	1.99	1.94	1.89	1.87	1.84	1.81	1.78	1.75	1.72
17	3.03	2.64	2.44	2.31	2.22	2.15	2.10	2.06	2.03	2.00	1.96	1.91	1.86	1.84	1.81	1.78	1.75	1.72	1.69
18	3.01	2.62	2.42	2.29	2.20	2.13	2.08	2.04	2.00	1.98	1.93	1.89	1.84	1.81	1.78	1.75	1.72	1.69	1.66
19	2.99	2.61	2.40	2.27	2.18	2.11	2.06	2.02	1.98	1.96	1.91	1.86	1.81	1.79	1.76	1.73	1.70	1.67	1.63
20	2.97	2.59	2.38	2.25	2.16	2.09	2.04	2.00	1.96	1.94	1.89	1.84	1.79	1.77	1.74	1.71	1.68	1.64	1.61
21	2.96	2.57	2.36	2.23	2.14	2.08	2.02	1.98	1.95	1.92	1.87	1.83	1.78	1.75	1.72	1.69	1.66	1.62	1.59
22	2.95	2.56	2.35	2.22	2.13	2.06	2.01	1.97	1.93	1.90	1.86	1.81	1.76	1.73	1.70	1.67	1.64	1.60	1.57
23	2.94	2.55	2.34	2.21	2.11	2.05	1.99	1.95	1.92	1.89	1.84	1.80	1.74	1.72	1.69	1.66	1.62	1.59	1.55
24	2.93	2.54	2.33	2.19	2.10	2.04	1.98	1.94	1.91	1.88	1.83	1.78	1.73	1.70	1.67	1.64	1.61	1.57	1.53
25	2.92	2.53	2.32	2.18	2.09	2.02	1.97	1.93	1.89	1.87	1.82	1.77	1.72	1.69	1.66	1.63	1.59	1.56	1.52
26	2.91	2.52	2.31	2.17	2.08	2.01	1.96	1.92	1.88	1.86	1.81	1.76	1.71	1.68	1.65	1.61	1.58	1.54	1.50
27	2.90	2.51	2.30	2.17	2.07	2.00	1.95	1.91	1.87	1.85	1.80	1.75	1.70	1.67	1.64	1.60	1.57	1.53	1.49
28	2.89	2.50	2.29	2.16	2.06	2.00	1.94	1.90	1.87	1.84	1.79	1.74	1.69	1.66	1.63	1.59	1.56	1.52	1.48
29	2.89	2.50	2.28	2.15	2.06	1.99	1.93	1.89	1.86	1.83	1.78	1.73	1.68	1.65	1.62	1.58	1.55	1.51	1.47
30	2.88	2.49	2.28	2.14	2.05	1.98	1.93	1.88	1.85	1.82	1.77	1.72	1.67	1.64	1.61	1.57	1.54	1.50	1.46
40	2.84	2.44	2.23	2.09	2.00	1.93	1.87	1.83	1.79	1.76	1.71	1.66	1.61	1.57	1.54	1.51	1.47	1.42	1.38
60	2.79	2.39	2.18	2.04	1.95	1.87	1.82	1.77	1.74	1.71	1.66	1.60	1.54	1.51	1.48	1.44	1.40	1.35	1.29
120	2.75	2.35	2.13	1.99	1.90	1.82	1.77	1.72	1.68	1.65	1.60	1.55	1.48	1.45	1.41	1.37	1.32	1.26	1.19
∞	2.71	2.30	2.08	1.94	1.85	1.77	1.72	1.67	1.63	1.60	1.55	1.49	1.42	1.38	1.34	1.30	1.24	1.17	1.00

$F_{.95}(n_1, n_2)$

n_1 = degrees of freedom for numerator

n_2 = degrees of freedom for denominator

n_2 \ n_1	1	2	3	4	5	6	7	8	9	10	12	15	20	24	30	40	60	120	∞
1	161.4	199.5	215.7	224.6	230.2	234.0	236.8	238.9	240.5	241.9	243.9	245.9	248.0	249.1	250.1	251.1	252.2	253.3	254.3
2	18.51	19.00	19.16	19.25	19.30	19.33	19.35	19.37	19.38	19.40	19.41	19.43	19.45	19.45	19.46	19.47	19.48	19.49	19.50
3	10.13	9.55	9.28	9.12	9.01	8.94	8.89	8.85	8.81	8.79	8.74	8.70	8.66	8.64	8.62	8.59	8.57	8.55	8.53
4	7.71	6.94	6.59	6.39	6.26	6.16	6.09	6.04	6.00	5.96	5.91	5.86	5.80	5.77	5.75	5.72	5.69	5.66	5.63
5	6.61	5.79	5.41	5.19	5.05	4.95	4.88	4.82	4.77	4.74	4.68	4.62	4.56	4.53	4.50	4.46	4.43	4.40	4.36
6	5.99	5.14	4.76	4.53	4.39	4.28	4.21	4.15	4.10	4.06	4.00	3.94	3.87	3.84	3.81	3.77	3.74	3.70	3.67
7	5.59	4.74	4.35	4.12	3.97	3.87	3.79	3.73	3.68	3.64	3.57	3.51	3.44	3.41	3.38	3.34	3.30	3.27	3.23
8	5.32	4.46	4.07	3.84	3.69	3.58	3.50	3.44	3.39	3.35	3.28	3.22	3.15	3.12	3.08	3.04	3.01	2.97	2.93
9	5.12	4.26	3.86	3.63	3.48	3.37	3.29	3.23	3.18	3.14	3.07	3.01	2.94	2.90	2.86	2.83	2.79	2.75	2.71
10	4.96	4.10	3.71	3.48	3.33	3.22	3.14	3.07	3.02	2.98	2.91	2.85	2.77	2.74	2.70	2.66	2.62	2.58	2.54
11	4.84	3.98	3.59	3.36	3.20	3.09	3.01	2.95	2.90	2.85	2.79	2.72	2.65	2.61	2.57	2.53	2.49	2.45	2.40
12	4.75	3.89	3.49	3.26	3.11	3.00	2.91	2.85	2.80	2.75	2.69	2.62	2.54	2.51	2.47	2.43	2.38	2.34	2.30
13	4.67	3.81	3.41	3.18	3.03	2.92	2.83	2.77	2.71	2.67	2.60	2.53	2.46	2.42	2.38	2.34	2.30	2.25	2.21
14	4.60	3.74	3.34	3.11	2.96	2.85	2.76	2.70	2.65	2.60	2.53	2.46	2.39	2.35	2.31	2.27	2.22	2.18	2.13
15	4.54	3.68	3.29	3.06	2.90	2.79	2.71	2.64	2.59	2.54	2.48	2.40	2.33	2.29	2.25	2.20	2.16	2.11	2.07
16	4.49	3.63	3.24	3.01	2.85	2.74	2.66	2.59	2.54	2.49	2.42	2.35	2.28	2.24	2.19	2.15	2.11	2.06	2.01
17	4.45	3.59	3.20	2.96	2.81	2.70	2.61	2.55	2.49	2.45	2.38	2.31	2.23	2.19	2.15	2.10	2.06	2.01	1.96
18	4.41	3.55	3.16	2.93	2.77	2.66	2.58	2.51	2.46	2.41	2.34	2.27	2.19	2.15	2.11	2.06	2.02	1.97	1.92
19	4.38	3.52	3.13	2.90	2.74	2.63	2.54	2.48	2.42	2.38	2.31	2.23	2.16	2.11	2.07	2.03	1.98	1.93	1.88
20	4.35	3.49	3.10	2.87	2.71	2.60	2.51	2.45	2.39	2.35	2.28	2.20	2.12	2.08	2.04	1.99	1.95	1.90	1.84
21	4.32	3.47	3.07	2.84	2.68	2.57	2.49	2.42	2.37	2.32	2.25	2.18	2.10	2.05	2.01	1.96	1.92	1.87	1.81
22	4.30	3.44	3.05	2.82	2.66	2.55	2.46	2.40	2.34	2.30	2.23	2.15	2.07	2.03	1.98	1.94	1.89	1.84	1.78
23	4.28	3.42	3.03	2.80	2.64	2.53	2.44	2.37	2.32	2.27	2.20	2.13	2.05	2.01	1.96	1.91	1.86	1.81	1.76
24	4.26	3.40	3.01	2.78	2.62	2.51	2.42	2.36	2.30	2.25	2.18	2.11	2.03	1.98	1.94	1.89	1.84	1.79	1.73
25	4.24	3.39	2.99	2.76	2.60	2.49	2.40	2.34	2.28	2.24	2.16	2.09	2.01	1.96	1.92	1.87	1.82	1.77	1.71
26	4.23	3.37	2.98	2.74	2.59	2.47	2.39	2.32	2.27	2.22	2.15	2.07	1.99	1.95	1.90	1.85	1.80	1.75	1.69
27	4.21	3.35	2.96	2.73	2.57	2.46	2.37	2.31	2.25	2.20	2.13	2.06	1.97	1.93	1.88	1.84	1.79	1.73	1.67
28	4.20	3.34	2.95	2.71	2.56	2.45	2.36	2.29	2.24	2.19	2.12	2.04	1.96	1.91	1.87	1.82	1.77	1.71	1.65
29	4.18	3.33	2.93	2.70	2.55	2.43	2.35	2.28	2.22	2.18	2.10	2.03	1.94	1.90	1.85	1.81	1.75	1.70	1.64
30	4.17	3.32	2.92	2.69	2.53	2.42	2.33	2.27	2.21	2.16	2.09	2.01	1.93	1.89	1.84	1.79	1.74	1.68	1.62
40	4.08	3.23	2.84	2.61	2.45	2.34	2.25	2.18	2.12	2.08	2.00	1.92	1.84	1.79	1.74	1.69	1.64	1.58	1.51
60	4.00	3.15	2.76	2.53	2.37	2.25	2.17	2.10	2.04	1.99	1.92	1.84	1.75	1.70	1.65	1.59	1.53	1.47	1.39
120	3.92	3.07	2.68	2.45	2.29	2.17	2.09	2.02	1.96	1.91	1.83	1.75	1.66	1.61	1.55	1.50	1.43	1.35	1.25
∞	3.84	3.00	2.60	2.37	2.21	2.10	2.01	1.94	1.88	1.83	1.75	1.67	1.57	1.52	1.46	1.39	1.32	1.22	1.00

Adapted with permission from *Biometrika Tables for Statisticians*, Vol. I (2d ed.), edited by E. S. Pearson and H. O. Hartley, Copyright 1958, Cambridge University Press.

Table 4 (continued).

$F_{.975}\,(n_1, n_2)$

n_1 = degrees of freedom for numerator

n_2 = degrees of freedom for denominator

n_2 \ n_1	1	2	3	4	5	6	7	8	9	10	12	15	20	24	30	40	60	120	∞
1	647.8	799.5	864.2	899.6	921.8	937.1	948.2	956.7	963.3	968.6	976.7	984.9	993.1	997.2	1001	1006	1010	1014	1018
2	38.51	39.00	39.17	39.25	39.30	39.33	39.36	39.37	39.39	39.40	39.41	39.43	39.45	39.46	39.46	39.47	39.48	39.49	39.50
3	17.44	16.04	15.44	15.10	14.88	14.73	14.62	14.54	14.47	14.42	14.34	14.25	14.17	14.12	14.08	14.04	13.99	13.95	13.90
4	12.22	10.65	9.98	9.60	9.36	9.20	9.07	8.98	8.90	8.84	8.75	8.66	8.56	8.51	8.46	8.41	8.36	8.31	8.26
5	10.01	8.43	7.76	7.39	7.15	6.98	6.85	6.76	6.68	6.62	6.52	6.43	6.33	6.28	6.23	6.18	6.12	6.07	6.02
6	8.81	7.26	6.60	6.23	5.99	5.82	5.70	5.60	5.52	5.46	5.37	5.27	5.17	5.12	5.07	5.01	4.96	4.90	4.85
7	8.07	6.54	5.89	5.52	5.29	5.12	4.99	4.90	4.82	4.76	4.67	4.57	4.47	4.42	4.36	4.31	4.25	4.20	4.14
8	7.57	6.06	5.42	5.05	4.82	4.65	4.53	4.43	4.36	4.30	4.20	4.10	4.00	3.95	3.89	3.84	3.78	3.73	3.67
9	7.21	5.71	5.08	4.72	4.48	4.32	4.20	4.10	4.03	3.96	3.87	3.77	3.67	3.61	3.56	3.51	3.45	3.39	3.33
10	6.94	5.46	4.83	4.47	4.24	4.07	3.95	3.85	3.78	3.72	3.62	3.52	3.42	3.37	3.31	3.26	3.20	3.14	3.08
11	6.72	5.26	4.63	4.28	4.04	3.88	3.76	3.66	3.59	3.53	3.43	3.33	3.23	3.17	3.12	3.06	3.00	2.94	2.88
12	6.55	5.10	4.47	4.12	3.89	3.73	3.61	3.51	3.44	3.37	3.28	3.18	3.07	3.02	2.96	2.91	2.85	2.79	2.72
13	6.41	4.97	4.35	4.00	3.77	3.60	3.48	3.39	3.31	3.25	3.15	3.05	2.95	2.89	2.84	2.78	2.72	2.66	2.60
14	6.30	4.86	4.24	3.89	3.66	3.50	3.38	3.29	3.21	3.15	3.05	2.95	2.84	2.79	2.73	2.67	2.61	2.55	2.49
15	6.20	4.77	4.15	3.80	3.58	3.41	3.29	3.20	3.12	3.06	2.96	2.86	2.76	2.70	2.64	2.59	2.52	2.46	2.40
16	6.12	4.69	4.08	3.73	3.50	3.34	3.22	3.12	3.05	2.99	2.89	2.79	2.68	2.63	2.57	2.51	2.45	2.38	2.32
17	6.04	4.62	4.01	3.66	3.44	3.28	3.16	3.06	2.98	2.92	2.82	2.72	2.62	2.56	2.50	2.44	2.38	2.32	2.25
18	5.98	4.56	3.95	3.61	3.38	3.22	3.10	3.01	2.93	2.87	2.77	2.67	2.56	2.50	2.44	2.38	2.32	2.26	2.19
19	5.92	4.51	3.90	3.56	3.33	3.17	3.05	2.96	2.88	2.82	2.72	2.62	2.51	2.45	2.39	2.33	2.27	2.20	2.13
20	5.87	4.46	3.86	3.51	3.29	3.13	3.01	2.91	2.84	2.77	2.68	2.57	2.46	2.41	2.35	2.29	2.22	2.16	2.09
21	5.83	4.42	3.82	3.48	3.25	3.09	2.97	2.87	2.80	2.73	2.64	2.53	2.42	2.37	2.31	2.25	2.18	2.11	2.04
22	5.79	4.38	3.78	3.44	3.22	3.05	2.93	2.84	2.76	2.70	2.60	2.50	2.39	2.33	2.27	2.21	2.14	2.08	2.00
23	5.75	4.35	3.75	3.41	3.18	3.02	2.90	2.81	2.73	2.67	2.57	2.47	2.36	2.30	2.24	2.18	2.11	2.04	1.97
24	5.72	4.32	3.72	3.38	3.15	2.99	2.87	2.78	2.70	2.64	2.54	2.44	2.33	2.27	2.21	2.15	2.08	2.01	1.94
25	5.69	4.29	3.69	3.35	3.13	2.97	2.85	2.75	2.68	2.61	2.51	2.41	2.30	2.24	2.18	2.12	2.05	1.98	1.91
26	5.66	4.27	3.67	3.33	3.10	2.94	2.82	2.73	2.65	2.59	2.49	2.39	2.28	2.22	2.16	2.09	2.03	1.95	1.88
27	5.63	4.24	3.65	3.31	3.08	2.92	2.80	2.71	2.63	2.57	2.47	2.36	2.25	2.19	2.13	2.07	2.00	1.93	1.85
28	5.61	4.22	3.63	3.29	3.06	2.90	2.78	2.69	2.61	2.55	2.45	2.34	2.23	2.17	2.11	2.05	1.98	1.91	1.83
29	5.59	4.20	3.61	3.27	3.04	2.88	2.76	2.67	2.59	2.53	2.43	2.32	2.21	2.15	2.09	2.03	1.96	1.89	1.81
30	5.57	4.18	3.59	3.25	3.03	2.87	2.75	2.65	2.57	2.51	2.41	2.31	2.20	2.14	2.07	2.01	1.94	1.87	1.79
40	5.42	4.05	3.46	3.13	2.90	2.74	2.62	2.53	2.45	2.39	2.29	2.18	2.07	2.01	1.94	1.88	1.80	1.72	1.64
60	5.29	3.93	3.34	3.01	2.79	2.63	2.51	2.41	2.33	2.27	2.17	2.06	1.94	1.88	1.82	1.74	1.67	1.58	1.48
120	5.15	3.80	3.23	2.89	2.67	2.52	2.39	2.30	2.22	2.16	2.05	1.94	1.82	1.76	1.69	1.61	1.53	1.43	1.31
∞	5.02	3.69	3.12	2.79	2.57	2.41	2.29	2.19	2.11	2.05	1.94	1.83	1.71	1.64	1.57	1.48	1.39	1.27	1.00

$F_{.99}(n_1, n_2)$

n_1 = degrees of freedom for numerator

n_2 = degrees of freedom for denominator

n_2 \ n_1	1	2	3	4	5	6	7	8	9	10	12	15	20	24	30	40	60	120	∞
1	4052	4999.5	5403	5625	5764	5859	5928	5982	6022	6056	6106	6157	6209	6235	6261	6287	6313	6339	6366
2	98.50	99.00	99.17	99.25	99.30	99.33	99.36	99.37	99.39	99.40	99.42	99.43	99.45	99.46	99.47	99.47	99.48	99.49	99.50
3	34.12	30.82	29.46	28.71	28.24	27.91	27.67	27.49	27.35	27.23	27.05	26.87	26.69	26.60	26.50	26.41	26.32	26.22	26.13
4	21.20	18.00	16.69	15.98	15.52	15.21	14.98	14.80	14.66	14.55	14.37	14.20	14.02	13.93	13.84	13.75	13.65	13.56	13.46
5	16.26	13.27	12.06	11.39	10.97	10.67	10.46	10.29	10.16	10.05	9.89	9.72	9.55	9.47	9.38	9.29	9.20	9.11	9.02
6	13.75	10.92	9.78	9.15	8.75	8.47	8.26	8.10	7.98	7.87	7.72	7.56	7.40	7.31	7.23	7.14	7.06	6.97	6.88
7	12.25	9.55	8.45	7.85	7.46	7.19	6.99	6.84	6.72	6.62	6.47	6.31	6.16	6.07	5.99	5.91	5.82	5.74	5.65
8	11.26	8.65	7.59	7.01	6.63	6.37	6.18	6.03	5.91	5.81	5.67	5.52	5.36	5.28	5.20	5.12	5.03	4.95	4.86
9	10.56	8.02	6.99	6.42	6.06	5.80	5.61	5.47	5.35	5.26	5.11	4.96	4.81	4.73	4.65	4.57	4.48	4.40	4.31
10	10.04	7.56	6.55	5.99	5.64	5.39	5.20	5.06	4.94	4.85	4.71	4.56	4.41	4.33	4.25	4.17	4.08	4.00	3.91
11	9.65	7.21	6.22	5.67	5.32	5.07	4.89	4.74	4.63	4.54	4.40	4.25	4.10	4.02	3.94	3.86	3.78	3.69	3.60
12	9.33	6.93	5.95	5.41	5.06	4.82	4.64	4.50	4.39	4.30	4.16	4.01	3.86	3.78	3.70	3.62	3.54	3.45	3.36
13	9.07	6.70	5.74	5.21	4.86	4.62	4.44	4.30	4.19	4.10	3.96	3.82	3.66	3.59	3.51	3.43	3.34	3.25	3.17
14	8.86	6.51	5.56	5.04	4.69	4.46	4.28	4.14	4.03	3.94	3.80	3.66	3.51	3.43	3.35	3.27	3.18	3.09	3.00
15	8.68	6.36	5.42	4.89	4.56	4.32	4.14	4.00	3.89	3.80	3.67	3.52	3.37	3.29	3.21	3.13	3.05	2.96	2.87
16	8.53	6.23	5.29	4.77	4.44	4.20	4.03	3.89	3.78	3.69	3.55	3.41	3.26	3.18	3.10	3.02	2.93	2.84	2.75
17	8.40	6.11	5.18	4.67	4.34	4.10	3.93	3.79	3.68	3.59	3.46	3.31	3.16	3.08	3.00	2.92	2.83	2.75	2.65
18	8.29	6.01	5.09	4.58	4.25	4.01	3.84	3.71	3.60	3.51	3.37	3.23	3.08	3.00	2.92	2.84	2.75	2.66	2.57
19	8.18	5.93	5.01	4.50	4.17	3.94	3.77	3.63	3.52	3.43	3.30	3.15	3.00	2.92	2.84	2.76	2.67	2.58	2.49
20	8.10	5.85	4.94	4.43	4.10	3.87	3.70	3.56	3.46	3.37	3.23	3.09	2.94	2.86	2.78	2.69	2.61	2.52	2.42
21	8.02	5.78	4.87	4.37	4.04	3.81	3.64	3.51	3.40	3.31	3.17	3.03	2.88	2.80	2.72	2.64	2.55	2.46	2.36
22	7.95	5.72	4.82	4.31	3.99	3.76	3.59	3.45	3.35	3.26	3.12	2.98	2.83	2.75	2.67	2.58	2.50	2.40	2.31
23	7.88	5.66	4.76	4.26	3.94	3.71	3.54	3.41	3.30	3.21	3.07	2.93	2.78	2.70	2.62	2.54	2.45	2.35	2.26
24	7.82	5.61	4.72	4.22	3.90	3.67	3.50	3.36	3.26	3.17	3.03	2.89	2.74	2.66	2.58	2.49	2.40	2.31	2.21
25	7.77	5.57	4.68	4.18	3.85	3.63	3.46	3.32	3.22	3.13	2.99	2.85	2.70	2.62	2.54	2.45	2.36	2.27	2.17
26	7.72	5.53	4.64	4.14	3.82	3.59	3.42	3.29	3.18	3.09	2.96	2.81	2.66	2.58	2.50	2.42	2.33	2.23	2.13
27	7.68	5.49	4.60	4.11	3.78	3.56	3.39	3.26	3.15	3.06	2.93	2.78	2.63	2.55	2.47	2.38	2.29	2.20	2.10
28	7.64	5.45	4.57	4.07	3.75	3.53	3.36	3.23	3.12	3.03	2.90	2.75	2.60	2.52	2.44	2.35	2.26	2.17	2.06
29	7.60	5.42	4.54	4.04	3.73	3.50	3.33	3.20	3.09	3.00	2.87	2.73	2.57	2.49	2.41	2.33	2.23	2.14	2.03
30	7.56	5.39	4.51	4.02	3.70	3.47	3.30	3.17	3.07	2.98	2.84	2.70	2.55	2.47	2.39	2.30	2.21	2.11	2.01
40	7.31	5.18	4.31	3.83	3.51	3.29	3.12	2.99	2.89	2.80	2.66	2.52	2.37	2.29	2.20	2.11	2.02	1.92	1.80
60	7.08	4.98	4.13	3.65	3.34	3.12	2.95	2.82	2.72	2.63	2.50	2.35	2.20	2.12	2.03	1.94	1.84	1.73	1.60
120	6.85	4.79	3.95	3.48	3.17	2.96	2.79	2.66	2.56	2.47	2.34	2.19	2.03	1.95	1.86	1.76	1.66	1.53	1.38
∞	6.63	4.61	3.78	3.32	3.02	2.80	2.64	2.51	2.41	2.32	2.18	2.04	1.88	1.79	1.70	1.59	1.47	1.32	1.00

Table 5 Percentiles of the x^2 Distribution.

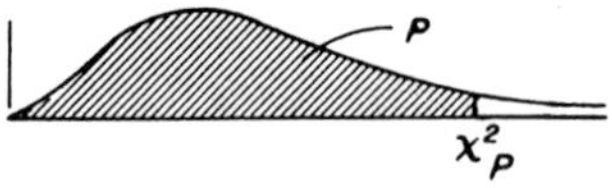

Values of χ^2_P corresponding to P

df	$\chi^2_{.005}$	$\chi^2_{.01}$	$\chi^2_{.025}$	$\chi^2_{.05}$	$\chi^2_{.10}$	$\chi^2_{.90}$	$\chi^2_{.95}$	$\chi^2_{.975}$	$\chi^2_{.99}$	$\chi^2_{.995}$
1	.000039	.00016	.00098	.0039	.0158	2.71	3.84	5.02	6.63	7.88
2	.0100	.0201	.0506	.1026	.2107	4.61	5.99	7.38	9.21	10.60
3	.0717	.115	.216	.352	.584	6.25	7.81	9.35	11.34	12.84
4	.207	.297	.484	.711	1.064	7.78	9.49	11.14	13.28	14.86
5	.412	.554	.831	1.15	1.61	9.24	11.07	12.83	15.09	16.75
6	.676	.872	1.24	1.64	2.20	10.64	12.59	14.45	16.81	18.55
7	.989	1.24	1.69	2.17	2.83	12.02	14.07	16.01	18.48	20.28
8	1.34	1.65	2.18	2.73	3.49	13.36	15.51	17.53	20.09	21.96
9	1.73	2.09	2.70	3.33	4.17	14.68	16.92	19.02	21.67	23.59
10	2.16	2.56	3.25	3.94	4.87	15.99	18.31	20.48	23.21	25.19
11	2.60	3.05	3.82	4.57	5.58	17.28	19.68	21.92	24.73	26.76
12	3.07	3.57	4.40	5.23	6.30	18.55	21.03	23.34	26.22	28.30
13	3.57	4.11	5.01	5.89	7.04	19.81	22.36	24.74	27.69	29.82
14	4.07	4.66	5.63	6.57	7.79	21.06	23.68	26.12	29.14	31.32
15	4.60	5.23	6.26	7.26	8.55	22.31	25.00	27.49	30.58	32.80
16	5.14	5.81	6.91	7.96	9.31	23.54	26.30	28.85	32.00	34.27
18	6.26	7.01	8.23	9.39	10.86	25.99	28.87	31.53	34.81	37.16
20	7.43	8.26	9.59	10.85	12.44	28.41	31.41	34.17	37.57	40.00
24	9.89	10.86	12.40	13.85	15.66	33.20	36.42	39.36	42.98	45.56
30	13.79	14.95	16.79	18.49	20.60	40.26	43.77	46.98	50.89	53.67
40	20.71	22.16	24.43	26.51	29.05	51.81	55.76	59.34	63.69	66.77
60	35.53	37.48	40.48	43.19	46.46	74.40	79.08	83.30	88.38	91.95
120	83.85	86.92	91.58	95.70	100.62	140.23	146.57	152.21	158.95	163.64

For large degrees of freedom,

$$\chi^2_P = \tfrac{1}{2}(z_P + \sqrt{2\nu - 1})^2 \text{ approximately,}$$

where ν = degrees of freedom and z_P is given in Table A-2.

Table 6 Factors for Control Charts for Variables X, Y, s, R: Normal Universe Factors for Computing Central Lines and 3-Sigma Control Limits

	Chart for Averages			Chart for Standard Deviations						Chart for Ranges						
Observations in	Factors for Control Limits			Factors for Central Line		Factors for Control Limits				Factors for Central Line		Factors for Control Limits				
Sample, n	A	A_2	A_3	c_4	$1/c_4$	B_3	B_4	B_5	B_6	d_2	$1/d_2$	d_3	D_1	D_2	D_3	D_4
2	2.121	1.880	2.659	0.7979	1.2533	0	3.267	0	2.606	1.128	0.8865	0.853	0	3.686	0	3.267
3	1.732	1.023	1.954	0.8862	1.1284	0	2.568	0	2.276	1.693	0.5907	0.888	0	4.358	0	2.574
4	1.500	0.729	1.628	0.9213	1.0854	0	2.266	0	2.088	2.059	0.4857	0.880	0	4.698	0	2.282
5	1.342	0.577	1.427	0.9400	1.0638	0	2.089	0	1.964	2.326	0.4299	0.864	0	4.918	0	2.114
6	1.225	0.483	1.287	0.9515	1.0510	0.030	1.970	0.029	1.874	2.534	0.3946	0.848	0	5.078	0	2.004
7	1.134	0.419	1.182	0.9594	1.0423	0.118	1.882	0.113	1.806	2.704	0.3698	0.833	0.204	5.204	0.076	1.924
8	1.061	0.373	1.099	0.9650	1.0363	0.185	1.815	0.179	1.751	2.847	0.3512	0.820	0.388	5.306	0.136	1.864
9	1.000	0.337	1.032	0.9693	1.0317	0.239	1.761	0.232	1.707	2.970	0.3367	0.808	0.547	5.393	0.184	1.816
10	0.949	0.308	0.975	0.9727	1.0281	0.284	1.716	0.276	1.669	3.078	0.3249	0.797	0.687	5.469	0.223	1.777
11	0.905	0.285	0.927	0.9754	1.0252	0.321	1.679	0.313	1.637	3.173	0.3152	0.787	0.811	5.535	0.256	1.744
12	0.866	0.266	0.886	0.9776	1.0229	0.354	1.646	0.346	1.610	3.258	0.3069	0.778	0.922	5.594	0.283	1.717
13	0.832	0.249	0.850	0.9794	1.0210	0.382	1.618	0.374	1.585	3.336	0.2998	0.770	1.025	5.647	0.307	1.693
14	0.802	0.235	0.817	0.9810	1.0194	0.406	1.594	0.399	1.563	3.407	0.2935	0.763	1.118	5.696	0.328	1.672
15	0.775	0.223	0.789	0.9823	1.0180	0.428	1.572	0.421	1.544	3.472	0.2880	0.756	1.203	5.741	0.347	1.653
16	0.750	0.212	0.763	0.9835	1.0168	0.448	1.552	0.440	1.526	3.532	0.2831	0.750	1.282	5.782	0.363	1.637
17	0.728	0.203	0.739	0.9845	1.0157	0.466	1.534	0.458	1.511	3.588	0.2787	0.744	1.356	5.820	0.378	1.622
18	0.707	0.194	0.718	0.9854	1.0148	0.482	1.518	0.475	1.496	3.640	0.2747	0.739	1.424	5.856	0.391	1.608
19	0.688	0.187	0.698	0.9862	1.0140	0.497	1.503	0.490	1.483	3.689	0.2711	0.734	1.487	5.891	0.403	1.597
20	0.671	0.180	0.680	0.9869	1.0133	0.510	1.490	0.504	1.470	3.735	0.2677	0.729	1.549	5.921	0.415	1.585
21	0.655	0.173	0.663	0.9876	1.0126	0.523	1.477	0.516	1.459	3.778	0.2647	0.724	1.605	5.951	0.425	1.575
22	0.640	0.167	0.647	0.9882	1.0119	0.534	1.466	0.528	1.448	3.819	0.2618	0.720	1.659	5.979	0.434	1.566
23	0.626	0.162	0.633	0.9887	1.0114	0.545	1.455	0.539	1.438	3.858	0.2592	0.716	1.710	6.006	0.443	1.557
24	0.612	0.157	0.619	0.9892	1.0109	0.555	1.445	0.549	1.429	3.895	0.2567	0.712	1.759	6.031	0.451	1.548
25	0.600	0.153	0.606	0.9896	1.0105	0.565	1.435	0.559	1.420	3.931	0.2544	0.708	1.806	6.056	0.459	1.541

Table reproduced from ASTM-STP 15D by kind permission of the American Society for Testing and Materials.

Table 7 Table for values of "y," to be used in calculating AOQL.

c	0	1	2
y	0.368	0.841	1.372
c	3	4	5
y	1.946	2.544	3.172
c	6	7	8
y	3.810	4.465	5.150
c	9	10	11
y	5.836	6.535	7.234

Index

About the Author

THOMAS PYZDEK is President of Quality America, Inc., a quality control consulting, training, and software firm in Tucson, Arizona. He is certified by the American Society for Quality Control (ASQC), of which he is a senior member, as both a quality engineer and a reliability engineer. Mr. Pyzdek's experience in quality includes over 20 years of work with both commercial and military contractors. He has worked on products that range from simple cans to complex guided missile systems. As a consultant, he has worked with dozens of companies, both large and small, to plan and implement quality control and quality improvement systems.

The author of two books, *An SPC Primer* and *The Certified Quality Engineering Examination Study Guide*, Mr. Pyzdek has also published numerous articles on quality control. He has delivered papers at national quality congresses. He received the B.A. degree (1974) in economics from the University of Nebraska at Omaha. A recipient of the Hughes Engineering Master's Fellowship, Mr. Pyzdek received the M.S. degree (1982) in industrial engineering from the University of Arizona in Tucson. In 1988 he served on the first Board of Examiners for the prestigious Malcolm Baldrige National Quality Award.